Springer Theses

Recognizing Outstanding Ph.D. Research

For further volumes:
http://www.springer.com/series/8790

Aims and Scope

The series "Springer Theses" brings together a selection of the very best Ph.D. theses from around the world and across the physical sciences. Nominated and endorsed by two recognized specialists, each published volume has been selected for its scientific excellence and the high impact of its contents for the pertinent field of research. For greater accessibility to non-specialists, the published versions include an extended introduction, as well as a foreword by the student's supervisor explaining the special relevance of the work for the field. As a whole, the series will provide a valuable resource both for newcomers to the research fields described, and for other scientists seeking detailed background information on special questions. Finally, it provides an accredited documentation of the valuable contributions made by today's younger generation of scientists.

Theses are accepted into the series by invited nomination only and must fulfill all of the following criteria

- They must be written in good English.
- The topic should fall within the confines of Chemistry, Physics, Earth Sciences and related interdisciplinary fields such as Materials, Nanoscience, Chemical Engineering, Complex Systems and Biophysics.
- The work reported in the thesis must represent a significant scientific advance.
- If the thesis includes previously published material, permission to reproduce this must be gained from the respective copyright holder.
- They must have been examined and passed during the 12 months prior to nomination.
- Each thesis should include a foreword by the supervisor outlining the significance of its content.
- The theses should have a clearly defined structure including an introduction accessible to scientists not expert in that particular field.

Lea Caminada

Study of the Inclusive Beauty Production at CMS and Construction and Commissioning of the CMS Pixel Barrel Detector

Doctoral Thesis accepted by
ETH Zurich, Switzerland

Springer

Author
Dr. Lea Caminada
ETH Zurich
Zurich, Switzerland

Supervisor
Prof. Dr. Felicitas Pauss
CERN
Geneva, Switzerland

ISSN 2190-5053
ISBN 978-3-642-24561-9
DOI 10.1007/978-3-642-24562-6
Springer Heidelberg New York Dordrecht London

e-ISSN 2190-5061
e-ISBN 978-3-642-24562-6

Library of Congress Control Number: 2012930907

© Springer-Verlag Berlin Heidelberg 2012

This work is subject to copyright. All rights are reserved by the Publisher, whether the whole or part of the material is concerned, specifically the rights of translation, reprinting, reuse of illustrations, recitation, broadcasting, reproduction on microfilms or in any other physical way, and transmission or information storage and retrieval, electronic adaptation, computer software, or by similar or dissimilar methodology now known or hereafter developed. Exempted from this legal reservation are brief excerpts in connection with reviews or scholarly analysis or material supplied specifically for the purpose of being entered and executed on a computer system, for exclusive use by the purchaser of the work. Duplication of this publication or parts thereof is permitted only under the provisions of the Copyright Law of the Publisher's location, in its current version, and permission for use must always be obtained from Springer. Permissions for use may be obtained through RightsLink at the Copyright Clearance Center. Violations are liable to prosecution under the respective Copyright Law.

The use of general descriptive names, registered names, trademarks, service marks, etc. in this publication does not imply, even in the absence of a specific statement, that such names are exempt from the relevant protective laws and regulations and therefore free for general use.

While the advice and information in this book are believed to be true and accurate at the date of publication, neither the authors nor the editors nor the publisher can accept any legal responsibility for any errors or omissions that may be made. The publisher makes no warranty, express or implied, with respect to the material contained herein.

Printed on acid-free paper

Springer is part of Springer Science+Business Media (www.springer.com)

Abstract

Beauty quarks are produced with a large cross section at a yet unreached center-of-mass energy at the Large Hadron Collider (LHC), enabling precision measurements to improve our understanding of heavy avor physics. Within this thesis a study of the inclusive b-quark production at the CMS experiment is presented. As a result an analysis strategy is proposed based on the reconstruction of muons produced in the semileptonic decay of b-quarks. The semileptonic decays are exploited in the kinematic range of muon transverse momentum $p_T > 6$ GeV and muon pseudorapidity $-2.1 < \eta < 2.1$.

The analysis is applied to data recorded by the CMS detector during the first months of high-energy collision data-taking in April and May 2010, corresponding to an integrated luminosity of $L = 8.1$ nb^{-1}. For the first time, the total b-quark production cross-section has been measured at a center-of-mass energy of $\sqrt{s} = 7$ TeV. The differential b-quark production cross-section as a function of muon transverse momentum and pseudorapidity is determined and compared to leading-order and next-to-leading-order QCD predictions.

The second part of the thesis focusses on the construction and commissioning of the CMS pixel barrel detector, the central part of the CMS detector with about 48 million readout channels. The CMS pixel detector allows to measure secondary vertices with high precision and thus plays a key role in the analysis of events with b-quarks.

The integration of the CMS pixel barrel detector has been accomplished within about two years before it was installed into the CMS detector in Summer 2008. The large effort in commissioning and calibration resulted in the successful and stable operation of the CMS pixel detector during cosmic and proton-proton collision data taking.

Supervisor's Foreword

On 30 March 2010, proton beams were for the first time brought into collision at CERNs Large Hadron Collider (LHC) at an unprecedented centre-of-mass-energy of 7 TeV, a factor 3.5 higher than achieved up to that date. The high centre-of-mass energy, combined with the high luminosity will make it possible to investigate in great detail the fundamental interactions of quarks and gluons and to perform many sensitive searches for new particles well beyond today's limits. However, a thorough understanding of Standard Model processes is necessary before entering the exploitation of possible new physics signatures. Beauty quarks play an especially important role in this respect, since b-jets are predicted to be a characteristic event signature for a variety of new physics processes at LHC energies.

The CMS experiment—one of the four large LHC experiments—is a general-purpose detector designed to optimally exploit the physics potential of the LHC. Located inside the superconducting solenoid, which provides a 3.8 Tesla field, are the hadronic and electromagnetic calorimeters as well as the tracking system. The latter is based on silicon pixels and silicon strip detectors, with a total silicon area of 210 m^2. A multi-layer muon system embedded in the return yoke outside the solenoid completes the CMS detector.

In order to extract the first measurement of the inclusive b-production cross section, Lea Caminada proposed an analysis strategy using muons from the semileptonic decay of b-quarks as an event signature. A key variable in this analysis is the measurement of the transverse momentum of the muon with respect to the jet direction ($p_\perp^{rel}$), which is larger in b-jets than other jets and is thus used as the discriminating variable. This particle jet is a result of b-quark fragmentation. Only charged tracks are used for the reconstruction of these jets, resulting in an efficient jet reconstruction. Furthermore, this approach is particularly important for low-energy jets and allows use of first data collected by CMS before a detailed understanding of systematic effects in the reconstruction of low-energy calorimetric jets is achieved.

Detailed Monte Carlo studies were needed to prepare for the data analysis, with particular attention being given to a reliable Monte Carlo prediction of the $p_\perp^{rel}$ variable. Appropriate techniques to validate the analysis method using a data-driven approach had to be developed, which also illustrates the challenges one faces at the LHC to perform this analysis.

Lea succeeded in measuring for the first time the inclusive b-quark production cross section at a centre-of-mass-energy of 7 TeV using 8.1 nb^{-1} of data collected by CMS during the first month of data-taking in April 2010. This result represents a remarkable achievement at this early stage of data-taking and analysis.

This physics result would not have been possible without a high performance pixel detector, which Lea also contributed to in all its phases: construction, commissioning and first operation. This detector, located closest to the beam pipe, allows one to reconstruct secondary vertices from heavy flavor and tau decays and to generate track seeds for the track reconstruction in the silicon-strip detector surrounding the pixel layers. Technologically very demanding requirements are imposed on the pixel detector: it has to operate in an environment of high track multiplicity (about 1000 charged tracks are produced every 25 ns at design values of the LHC) and heavy irradiation. The barrel pixel detector was designed, constructed and tested at the Paul Scherrer Institute (PSI) in Switzerland, in a collaboration between PSI, ETH Zurich and the University of Zurich, under the leadership of Prof. Roland Horisberger from PSI.

Lea's work describes the overall pixel detector concept and provides detailed information about the detector modules, including their main building blocks and the readout and control systems. Lea has developed the software algorithms required for testing, calibration and monitoring all detector components during the integration at PSI, and later, during installation in the CMS pit at CERN. Each step in the construction and commissioning is very clearly explained and well illustrated by the corresponding results obtained. The reader can thus appreciate the different challenges that had to be faced in this cutting-edge technology and the importance of Lea's contributions, which ensured the very successful operation of the pixel detector, with less than 1% dead channels.

In addition to their impact on the pixel operation, the results of the work described in this thesis also set the stage for new analysis techniques using the CMS detector. The results also provide an excellent example of the close link between the development of forefront technologies and major progress in fundamental science, by exploiting the new energy frontier offered by the LHC.

Zurich, November 2011   Dr. Felicitas Pauss

Acknowledgments

It was a great pleasure and a unique experience to work in a large international collaboration like CMS. This work would not have been possible without the help of many people and I am grateful to everybody who contributed to its successful completion. However, there are some people I would like to especially mention.

I wish to express my sincere thank to Felicitas Pauss for giving me the opportunity to do my thesis under her supervision. I very much appreciated her competent advise, her enthusiasm and her continuos support.

I am grateful to Ralph Eichler for making it possible to carry out my PhD work within the CMS groups at ETH and PSI. I am pleased that Ralph Eichler readily accepted to be a co-examiner for this thesis and I would like to thank him for his valuable comments.

Special thanks go to Roland Horisberger for introducing me to the CMS pixel community and sharing his great expertise. Working in his group was a pleasant and challenging activity. I am happy that Roland Horisberger acted as a co-examiner for this work.

I enjoyed the supervision of Christoph Grab and I would like to thank him for his advise throughout all parts of this thesis. His suggestions improved this work considerably.

I am indebted to Wolfram Erdmann for his outstanding support with regard to every aspect of my work. He was always there when a question occurred and I could profit a lot from his expert knowledge. His excellent guidance truly enriched my research work.

I would like to thank my colleagues of the Institute of Particle Physics at ETH, the CMS Pixel Group at PSI and the CMS Collaboration. I was very lucky to meet talented, inspiring and interesting people.

Finally, I would like to sincerely thank my parents, my sister and Christian for their incessant support. Their patience, their advice and encouragement are invaluable for me.

Contents

1	**Introduction**		1
	References		4
2	**The CMS Experiment at the LHC**		7
	2.1	The Large Hadron Collider	7
	2.2	The CMS Detector	8
		2.2.1 Coordinate Conventions	9
		2.2.2 Solenoid	10
		2.2.3 Tracking Detectors	11
		2.2.4 Track Reconstruction	12
		2.2.5 Electromagnetic Calorimeter	13
		2.2.6 Hadronic Calorimeter	16
		2.2.7 Muon System	17
		2.2.8 Muon Reconstruction	19
		2.2.9 Trigger System	20
	References		22

Part I Study of the Inclusive b quark Production at CMS

3	**Heavy Flavor Physics**		25
	3.1	Quantum Chromodynamics	25
	3.2	Hadronic Collisions	27
		3.2.1 Event Kinematics	27
		3.2.2 Factorization	28
		3.2.3 Evolution of Parton Distribution Functions	30
	3.3	Heavy Quark Production	31
	3.4	The Fragmentation of Heavy Quarks	33
	3.5	Semileptonic Decays of Heavy Quarks	34

3.6	Monte Carlo Event Generators.	36
References		38

4 Study of the Inclusive Beauty Production ... 41
4.1	Concept	41
4.2	Event Simulation	42
4.3	Trigger	44
4.4	Jet Reconstruction	45
4.5	Event Selection	50
4.6	Signal Extraction	52
	4.6.1 Fitting Procedure	53
	4.6.2 Performance of the Fit	55
4.7	Validation of MC Templates	56
	4.7.1 Signal.	56
	4.7.2 Background.	58
4.8	Cross Section Measurement.	61
	4.8.1 Inclusive Cross Section.	61
	4.8.2 Differential Cross Section	62
4.9	Systematic Uncertainties.	63
	4.9.1 Trigger	64
	4.9.2 Tracking Efficiency and Misalignment	64
	4.9.3 Background Composition	65
	4.9.4 Fragmentation and Decay	66
	4.9.5 Production Mechanism	67
	4.9.6 Description of the Underlying Event	67
	4.9.7 Monte Carlo Statistics	68
	4.9.8 Luminosity	69
	4.9.9 Summary	69
4.10	Results	69
References		72

5 Results of First Collisions at $\sqrt{s} = 900$ GeV and $\sqrt{s} = 2.36$ TeV ... 75
5.1	Event Selection	75
5.2	Event Simulation	76
5.3	Muon Distributions	76
5.4	Track Distributions.	78
5.5	TrackJet Distributions.	79
5.6	$p_{\perp}^{\mathrm{rel}}$ Distribution	81
5.7	Conclusions.	83
References		84

6 Preliminary Results of First Collisions at $\sqrt{s} = 7$ TeV ... 85
6.1	Event Simulation	85

6.2		Event Selection		85
	6.2.1	Run Selection		86
	6.2.2	Trigger Selection		87
	6.2.3	Offline Selection		87
6.3		Signal Extraction		89
	6.3.1	Data-Driven Determination of Light Quark Background		89
	6.3.2	Data-driven Validation of $p_\perp^{rel}$ Templates in Signal Events		91
	6.3.3	$p_\perp^{rel}$ Fit		91
6.4		Results		93
6.5		Systematic Uncertainties		95
6.6		Conclusions		96
References				97

Part II Construction and Commissioning of the CMS Pixel Barrel Detector

7 The CMS Pixel Barrel Detector 101

7.1	Design of the CMS Pixel Barrel Detector		101
7.2	Detector Modules		102
	7.2.1	Sensor	103
	7.2.2	Readout Chip	103
	7.2.3	Token Bit Manager	105
7.3	Readout and Control System		105
	7.3.1	Analog Chain	106
	7.3.2	Front End Driver	107
	7.3.3	Supply Tube	108
	7.3.4	Communication and Control Unit	109
	7.3.5	Front End Controller	110
References			110

8 Construction and Commissioning of the CMS Pixel Barrel Detector 113

8.1	Low Level Hardware Testing Procedure		113
	8.1.1	Test Setup	114
	8.1.2	Software Architecture	114
	8.1.3	Testing Sequence	115
8.2	Performance Tests and Calibrations		119
8.3	Construction		121
	8.3.1	Module Mounting	122
	8.3.2	Supply Tube Assembly	124
	8.3.3	Integration of the Complete System	127

8.4		Installation into CMS	128
8.5		Commissioning and Performance	129
	8.5.1	Performance of the Optical Links	130
	8.5.2	Detector Module Functionality	131
	8.5.3	Results of the Detector Calibration	132
	8.5.4	Results of the Cosmic Run	132
8.6		Summary	135
References			135

9 Conclusion and Outlook ... 137

Appendix A: Maximum Likelihood Fits ... 139

Appendix B: Systematic Uncertainties ... 151

Appendix C: Preliminary Results of First Collision at $\sqrt{s} = 7$ TeV ... 153

Curriculum Vitae ... 155

Chapter 1
Introduction

The understanding of the matter that surrounds us has been intriguing scientists and philosophers ever since. The idea that all matter is composed of fundamental building blocks was first conceived of by Greek philosophers more than two thousand years ago. This idea remained untested until the early twentieth century when the first experiments investigating the subatomic structures were performed. A tremendous technological progress in the second half of the century allowed to develop new experimental methods and revolutionized the field of particle physics. The construction of large scale particle accelerators and ever more sophisticated detectors paved the way to a host of discoveries. Based on these experiments a new level of understanding has been gained. Nearly everything we currently know about the constituents of matter and their interactions can be described by a relativistic quantum field theory known as the Standard Model (SM) of elementary particle physics.

Since its introduction in the early 1970's by Glashow [1], Weinberg [2] and Salam [3], it has successfully explained a wide range of experimental results and precisely predicted a large variety of phenomena ranging from the elementary world to the realm of the universe. The interested reader is referred to [4–7] for a comprehensible introduction to the SM.

In the SM, matter is built from fermions (spin-$\frac{1}{2}$ particles) and the interaction between matter particles are described by the exchange of bosons (spin-1 particles). Three of the four fundamental forces, namely electromagnetism, weak interaction and strong interaction, are included in the SM. The gravitational force which is dominant at macroscopic distances is omitted since its effects are small on microscopic scales. There are 12 fermions divided into 6 quark flavors (u, d, s, c, b, t) and 6 lepton flavors ($e, \mu, \tau, \nu_e, \nu_\mu, \nu_\tau$), all of which possess charge conjugate states, called antiparticles. Moreover, the quarks and leptons are classified into three generations with very different mass scales ranging in case of the quarks from a few MeV for up and down quarks to about 171 GeV for the top quark.

All quarks and leptons are subject to the weak force, while only quarks and charged leptons (e, μ, τ) undergo electromagnetic interactions. A major achievement of the SM is the unification of the electromagnetic and the weak force into an electroweak

---

L. Caminada, *Study of the Inclusive Beauty Production at CMS and Construction and Commissioning of the CMS Pixel Barrel Detector*, Springer Theses, DOI: 10.1007/978-3-642-24562-6_1, © Springer-Verlag Berlin Heidelberg 2012

force embedded in the theory of Quantum Electrodynamics (QED). Therein the photon is the gauge boson of the electromagnetism and the W- and Z-bosons are the mediators of the weak interaction. The strong force is responsible for the interaction between color charged quarks and gluons where the latter are the force carriers in the theory of Quantum Chromodynamics (QCD). Excellent reading on the concepts of quantum field theories can be found in [8].

It should be emphasized that seven ($c, b, t, \nu_\tau, W/Z$, gluon) out of the 16 particles have been predicted by the SM before they have been observed experimentally. The SM is completed by the introduction of an additional particle called the Higgs boson. The Higgs boson plays an important role in the SM as it provides an explanation for the masses of the elementary particles and gives rise to the phenomenon of electroweak symmetry breaking. Despite the large effort, the experimental verification of the existence of the Higgs boson has not been crowned with success so far.

The search for the elusive Higgs particle has been one of the main motivations for the construction of the Large Hadron Collider (LHC) at CERN. The LHC is a proton-proton collider which has been designed to operate at a center-of-mass energy of $\sqrt{s} = 14\,\text{TeV}$. LHC operation started on March 30, 2010 and the first collisions happened at a center-of-mass energy of $\sqrt{s} = 7\,\text{TeV}$. It is foreseen to collect a significant amount of data at this lower energy before the collision energy will be increased to the design value in 2013. The energy of $\sqrt{s} = 14\,\text{TeV}$ is high enough to produce the Higgs boson and access the yet unexplored mass range up to about 1 TeV. The experiments at the LHC are supposed to finally answer one of the most urgent questions of modern particle physics and to either detect the Higgs boson and confirm the SM prediction or to exclude its existence and falsify this part of the SM. Readers interested in the design and the operation of the LHC are referred to [9], while more information on the LHC experiments is presented in [10].

In spite of the outstanding achievement of the SM in describing various aspects of the physics within its domain, it is nevertheless an incomplete theory. First of all, the theoretical framework is known to be no longer self-consistent when approaching energies of a few TeV. Furthermore, the SM leaves many fundamental questions unanswered, concerning for instance the nature of dark matter or the origin of the asymmetry between matter and antimatter in our universe. In an endeavor to provide explanations for these questions, many promising theories extending the SM have been developed. Supersymmetric theories or theories with extra-dimensions are the most prominent examples. Many of these theories predict physics effects which should be accessible at the energy scale available at the LHC. Hopes are high for discovering new phenomena at the LHC which would challenge the SM and possibly give crucial information about physics beyond the SM.

Owing to the need of good statistics high quality data for the search of the Higgs boson, the potential discovery will take its time. The short term goal of the CMS physics program is therefore the experimental confirmation of the SM at the higher energy scale the LHC is capable of providing.

Within the thesis presented here the production of beauty quarks at the LHC is studied at the CMS experiment. Due to the large b-quark production cross-section at the LHC, high statistics data samples are available soon after the LHC startup which renders the CMS experiment an excellent facility for the study of heavy flavor physics.

The b-quark is a third generation quark which was discovered in 1977 by the E288 experiment at Fermilab [11]. The large mass of the b-quark large mass (around 4.2 GeV) and its long lifetime provide a distinctive signature which allows to identify the particle experimentally. The production of heavy quarks is expected to be accurately described by perturbative QCD. The investigation of events containing beauty quarks is interesting as it allows to probe the predictions of the theory of strong interactions at the energy scale provided by the LHC. Furthermore, the production of b quarks is a major source of background for many searches to be performed at the LHC and therefore has to be well understood.

A measurement of the beauty quark production cross-section based on the semileptonic decay of b quarks into muons is performed in this thesis. Because of the large mass of the b quark, muons from semileptonic b-decays have larger transverse momenta with respect to the quark direction than muons from the decay of lighter quarks. In the experiment, the quark direction is approximated by the axis of the fragmentation jet and the transverse momentum of the muon relative to that axis ($p_\perp^{\mathrm{rel}}$) is measured. The contribution of b-events to the measured distribution is determined by performing a fit based on simulated template distributions for signal and background events.

The method of discriminating b-events from background events by means of the $p_\perp^{\mathrm{rel}}$ variable has been used for the first time by the UA1 experiment at the CERN SPS collider [12]. Since then it has become a well-established technique for the identification of beauty quarks: the $p_\perp^{\mathrm{rel}}$ method has been further applied for measuring the b-quark production cross-section at electron-positron [13, 14], electron-proton [15] and hadron-hadron colliders [16, 17].

Measurements of the b-quark production cross-section are available from collider experiments at different center-of-mass energies. A lot of progress has been made in understanding the b-quark production process [18], so that the measurements are meanwhile in reasonable agreement with the theoretical predictions in most regions of the phase space. Nonetheless, there is a great interest in verifying these results at the higher center-of-mass energy of the LHC, since the theoretical uncertainties are still sizeable. A first measurement of the b-quark production cross-section using the very early data from the LHC is presented within this work.

Even though the CMS detector is primarily designed for high transverse momentum physics, it is very well suited for heavy flavor physics thanks to the muon system with the potential to identify low transverse momentum muons and the excellent tracking detectors. In particular, CMS features a novel three-layer silicon pixel detector which allows for a precise and efficient reconstruction of secondary vertices from heavy flavor decays.

The barrel part of the CMS pixel detector was developed, designed and built at PSI in cooperation with ETH Zürich and the University of Zürich. In the framework of this thesis important contributions to the construction and commissioning of the pixel detector were made. This includes the assembly of the complete system together with

4 1 Introduction

the control and data acquisition system at PSI, the construction of the final system and the development of software algorithms needed for testing and calibration. The CMS pixel detector was installed into the CMS detector in summer 2008 and after a period of commissioning and calibration successfully operated during both cosmic and collision data taking.

The thesis is organized as follows: in Chap. 2 an overview of the experimental facility at the LHC and the CMS detector is presented. The theoretical framework important for the physics analysis described herein is introduced in Chap. 3. This includes a review of the main ideas of Quantum Chromodynamics and its application to hadronic collisions, an introduction to the physics of heavy quarks and an overview of the Monte Carlo event generators used in the analysis. Chapter 4 is devoted to the study of the inclusive b-quark production cross-section at CMS. The prospects for the measurement were determined based on simulated data. Furthermore, a detailed study of the main systematic effects of the measurement is accomplished. In December 2009, a first commissioning run producing proton-proton collisions at center -of-mass energies of $\sqrt{s} = 900$ GeV and $\sqrt{s} = 2.36$ TeV took place at the LHC. The data collected during this run was analyzed within this work and the results relevant for the study of the beauty production are summarized in Chap. 5. In Chap. 6 the preliminary results of the measurement of the inclusive b-quark production cross-section at a center-of-mass energy of $\sqrt{s} = 7$ TeV are presented. The measurement is based on the collision data recorded by the CMS experiment during the first months of high-energy collision data-taking in April and May 2010.

The second part of the thesis concentrates on the hardware related work performed within the CMS pixel detector group at PSI. In Chap. 7 the main aspects of the design and the functionality of the CMS pixel barrel detector are addressed, while Chap. 8 focusses on how this work contributed to the construction and commissioning of the detector.

Throughout this thesis, natural units are used in which $c = \hbar = 1$.

References

1. S. Glashow, Partial symmetries of weak interactions. Nucl. Phys. **22**, 579–588 (1961)
2. S. Weinberg, A model of leptons. Phys. Rev. Lett. **19**, 1264–1266 (1967)
3. A. Salam, Elementary particle physics: relativistic groups and analyticity, in *Nobel Symposium*, No. 8, ed. by N. Svartholm, vol. 367 (Almqvist and Wiksills, Stockholm, 1968)
4. D.H. Perkins, *Introduction to High Energy Physics*, 4th edn (Cambridge University Press, Cambridge, 2000)
5. B. Povh, K. Rith, C. Scholz, F. Zetsche, *Particles and Nuclei: An introduction to the Physical Concepts* (Springer, Berlin, 2003)
6. W.S.C. Williams, *Nuclear and Particle Physics* (Oxford University Press, Oxford, 1994)
7. F. Halzen, A.D. Martin, *Quarks and Leptons: An Introductory Course in Modern Particle Physics* (Wiley, New York, 1984)

References

8. M.E. Peskin, D.V. Schroeder, *An Introduction to Quantum Field Theory* (Addison-Wesley Advanced Book Program, Reading, 1995)
9. L. Evans, *The Large Hadron Collider: A Marvel of Technology* (EPFL Press, Lausanne, 2009)
10. D. Green, *At the Leading Edge: The ATLAS and CMS LHC Experiments* (World Scientific, New York, 2011)
11. S.W. Herb et al., Observation of a Dimuon resonance at 9.5 Gev in 400-GeV proton-nucleus collisions. Phys. Rev. Lett. **39**, 252–255 (1977)
12. UA1 Collaboration, Beauty production at the CERN proton-antiproton collider. Phys. Lett. B **186**, 237 (1987)
13. ALEPH Collaboration, Nucl. Instrum. Meth. A **346**, 461 (1994)
14. L3 Collaboration, Phys. Rev. Lett. B **252**, 703 (1990)
15. U. Langenegger, A Measurement of the Beauty and Charm Production Cross Sections at the ep collider HERA, Ph.D. Thesis, ETH Zürich, ETH No. 12676, 1998
16. DØ Collaboration, Inclusive μ and b-quark production cross sections in $p\overline{p}$ collisions at $\sqrt{s} = 1.8$ TeV. Phys. Rev. Lett. **74**, 3548 (1995)
17. CDF Collaboration, Measurement of the bottom quark production cross section using semi-leptonic decay electrons in $p\overline{p}$ collisions at $\sqrt{s} = 1.8$ TeV. Phys. Rev. Lett. **71**, 500–504 (1993)
18. M.L. Mangano, The saga of bottom production in $p\overline{p}$ collisions. AIP Conf. Proc. **753**, 247–260 (2005)

Chapter 2
The CMS Experiment at the LHC

The Large Hadron Collider (LHC) [1] is designed to collide proton beams at a center-of-mass energy of $\sqrt{s} = 14\,\mathrm{TeV}$ and a nominal instantaneous luminosity of $\mathcal{L} = 10^{34}\,\mathrm{cm}^{-2}\mathrm{s}^{-1}$. This represents a seven-fold increase in energy and a hundred-fold increase in integrated luminosity over the previous hadron collider experiments. The beam energy and the design luminosity have been chosen in order to study physics at the TeV energy scale. The main motivation of the LHC is to reveal the nature of electroweak symmetry breaking and to investigate potential manifestations of new physics phenomena beyond the SM.

The unprecedented high center-of-mass energy and luminosity at the LHC lead to a number of substantial experimental challenges. In this chapter an introduction of the experimental facility is presented. An overview of the basic aspects of the LHC is followed by a detailed description of the CMS experiment and its main subdetectors.

2.1 The Large Hadron Collider

The LHC accelerator is located at CERN near Geneva in the already existing LEP tunnel which has a length of 26.7 km. Four detectors are installed in the experimental caverns around the collision points: Two general purpose experiments, ATLAS [2] and CMS [3, 4], the LHCb [5] experiment dedicated to B Physics and the ALICE [6] experiment where the physics of heavy ion collisions is investigated.

To supply the LHC with pre-accelerated protons the existing CERN facilities have been upgraded. A schematic view of the LHC accelerator with the injection chain is shown in Fig. 2.1. The protons coming from the Super Proton Synchrotron (SPS) with an energy of 450 GeV are injected into the LHC, where they will be accelerated to an energy of 7 TeV in bunches with a nominal number of $1.15 \cdot 10^{11}$ particles per bunch. Superconducting dipole magnets which provide a magnetic field of 8.3 T are needed to keep the protons on the orbit during the acceleration. The superconducting magnets are cooled using liquid helium at a temperature of 1.9 K.

L. Caminada, *Study of the Inclusive Beauty Production at CMS and Construction and Commissioning of the CMS Pixel Barrel Detector*, Springer Theses, DOI: 10.1007/978-3-642-24562-6_2, © Springer-Verlag Berlin Heidelberg 2012

At the interaction point where the CMS experiment is located collisions happen every 25 ns, corresponding to a bunch crossing frequency of 40 MHz. The total proton-proton cross section at $\sqrt{s} = 14$ TeV is expected to be about 100 mb. Therefore, approximately 10^9 inelastic events per second will be observed in the multi-purpose experiments at design luminosity.

In the first year of LHC running the proton-proton collisions happen at a lower center-of-mass energy of $\sqrt{s} = 7$ TeV and at a lower luminosity than what is foreseen in the original design. A commissioning run with collisions at center-of-mass energies of $\sqrt{s} = 900$ GeV and 2.36 TeV took place in November and December 2009. The LHC operation at $\sqrt{s} = 7$ TeV started on March 30, 2010.

2.2 The CMS Detector

The CMS detector is a general purpose detector installed 100 m underground at the LHC interaction point 5 (P5) near the village of Cessy in France. The design of the CMS detector is driven by the challenges of a physics experiment in the LHC environment. Many of the physics benchmark channels have a small cross section and the background from QCD jet production is overwhelmingly dominant. A high rejection power with an optimal efficiency for rare channels has to be achieved. The reconstruction of lepton signatures is essential for the extraction of rare processes and an excellent muon and electron identification and momentum resolution is desired. Moreover, a precise measurement of secondary vertices and impact parameters is necessary for an efficient identification of heavy flavor and τ-lepton decays.

The short bunch crossing separation and the high event rate at the LHC impose further challenges to the design. At design luminosity, 23 inelastic interaction per bunch crossing will occur on average. This phenomenon is know as pile-up. The effect of pile-up can be reduced by using high-granularity detectors resulting in low occupancy. This requires a large number of detector channels and an excellent synchronization among them. In addition, a good time resolution is needed to discriminate the interaction under study from the interactions occurring in neighboring bunch crossings. Another difficulty arises from the large flux of particles near the interaction point which leads to high radiation levels and the need of radiation hard detectors and front-end electronics.

The CMS detector is divided into a silicon tracking system, an electromagnetic and a hadronic calorimeter and a muon system. A magnetic field of 3.8 T is provided by a superconducting solenoid magnet. The CMS detector is 22 m long, has a diameter of 15 m and an overall weight of 12.5 t. Figure 2.2 presents a schematic view of the CMS detector.

2.2 The CMS Detector

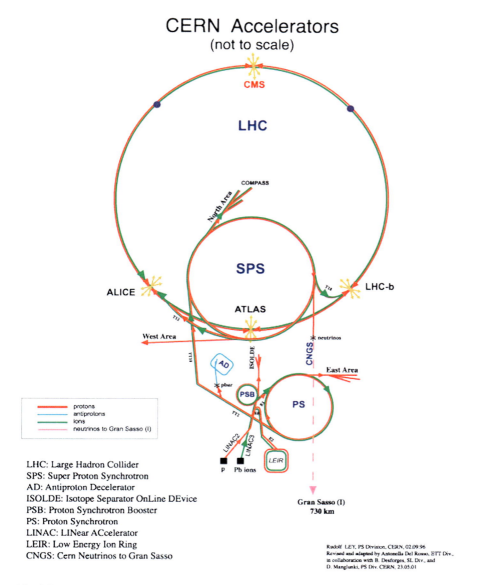

Fig. 2.1 The CERN accelerator complex. Schematic view of the LHC and the location of the four main experiments ATLAS, CMS, LHCb and ALICE

2.2.1 Coordinate Conventions

The CMS coordinate system is oriented such that the x-axis points south to the center of the LHC ring, the y-axis points vertically upward and the z-axis is in the direction

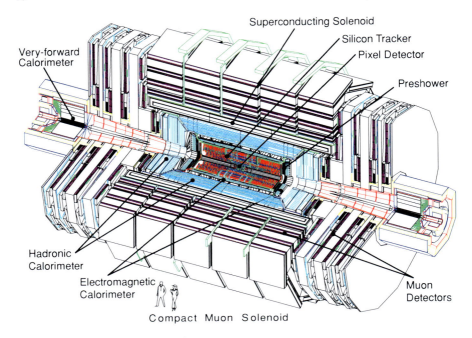

Fig. 2.2 Schematic view of the CMS detector

of the beam to the west. The azimuthal angle ϕ is measured from the x-axis in the xy plane and the radial coordinate in this plane is denoted by r. The polar angle θ is defined in the rz plane and the pseudorapidity is $\eta = -\ln\tan(\theta/2)$. The momentum component transverse to the beam direction, denoted by p_T, is computed from the x- and y-components, while the transverse energy is defined as $E_T = E\sin\theta$.

2.2.2 Solenoid

A superconducting solenoid magnet with a maximum magnetic field of 3.8 T provides the large bending power for high-energy charged particles to precisely measure their momentum in the tracking detectors. The magnetic coil is 13 m long, has an inner diameter of 6 m and accommodates the tracking and part of the calorimeter detectors. With these dimensions the CMS solenoid is the largest superconducting magnet ever built and has the capacity to store an energy of 2.6 GJ at full current. The magnetic flux is returned through a 10,000 t iron yoke in which the muon detector chambers are integrated. The CMS solenoid was fully tested and commissioned at the experimental surface hall during autumn 2006.

2.2 The CMS Detector

2.2.3 Tracking Detectors

CMS features an all silicon tracker with a total active area of $200\,m^2$. The tracking detector is divided into a pixel detector close to the interaction region and a strip detector covering radii between 0.2 and 1.2 m. At LHC design luminosity more than 1,000 particles are expected to traverse the tracking volume in each bunch crossing. This leads to a hit rate density of $1\,MHz/mm^2$ at a radius of 4 cm which imposes severe challenges to the design of the tracking detectors. With a pixel size of $100 \times 150\,\mu m^2$ an occupancy of less than 10^{-4} can be maintained in the pixel detector. As the particle flux decreases with the distance from the interaction point, sensors with a length of 10 cm and a pitch of $80\,\mu m$ can be used at intermediate radii (20–55 cm) and sensors with a length of 25 cm and a pitch of $180\,\mu m$ at the outermost radii (55–110 cm) with an occupancy of less than 3%. The sensor thickness is $285\,\mu m$ for the pixel and 320 or $500\,\mu m$ for the strip modules at the intermediate and outer radii, respectively. The thicker sensors for the outer tracking region permit to preserve a signal to noise ratio well above 10 in spite of the higher noise due to the larger capacity of the longer strips. To mitigate the radiation damage effects and prolong the lifetime of the detector modules, the tracking detectors are designed to run at subzero temperatures. The cooling is established using a mono-phase liquid cooling system whit C_6F_{14} as cooling fluid.

The pixel detector is built from 3 barrel layers at radii of 4.4, 7.3 and 10.2 cm and two end disks on each side at a distance of $z = \pm 34.5, \pm 46.5$ cm from the interaction point. It consists of 1,440 segmented silicon sensor modules with a total of 66 million readout channels. For each pixel the analog pulse height information is detected and read out.

The sensor surface of the barrel modules is parallel to the magnetic field, while the modules in the forward detector are tilted by $20°$. The charge carriers produced by a particle traversing the sensor thus experience the Lorentz force and do not drift along the electric field line anymore. Hence, the charge carriers are distributed over several pixels. The analog pulse height information can be used to calculate a center of gravity of the charge distribution which improves the hit information. In this way a position resolution of about $15\,\mu m$ in both the $r\phi$ and z directions is obtained (compare to Fig. 2.6). A detailed description of the design and the functioning of the CMS pixel barrel detector is given in Chap. 7.

The silicon strip tracker has a length of 5.8 m and a diameter of 2.4 m and is composed of four subsystems: the four-layer Tracker Inner Barrel (TIB), the six-layer tracker outer barrel (TOB) and on each side three-disk Tracker Inner Disks (TID) and nine-disk Tracker Endcaps (TEC). An rz-view of the tracker geometry is shown in Fig. 2.3.

The silicon strip tracker is built from 15,148 single-sided modules that provide 9.3 million readout channels. Modules for the TIB, the TID and the first four rings of the TEC are single-sided while the TOB and the outer three rings of the TEC are equipped with double-sided modules. A double-sided module is constructed from two single-sided modules glued back-to-back at a stereo angle of 100 mrad.

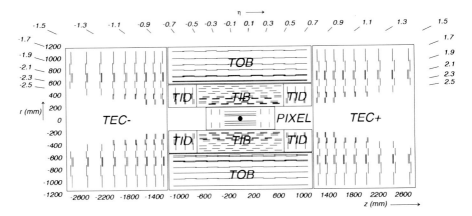

Fig. 2.3 rz-view of the CMS tracking detectors [4]. *Single lines* represent layers of modules equipped with one sensor, *double lines* indicate layers with back-to-back modules

2.2.4 Track Reconstruction

Due to the magnetic field charged particles travel through the tracking detectors on a helix trajectory which is described by 5 parameters: the curvature κ, the track azimuthal angle ϕ and polar angle η, the signed transverse impact parameter d_0 and the longitudinal impact parameter z_0. The transverse (longitudinal) impact parameter of a track is defined as the transverse (longitudinal) distance of closest approach of the track to the primary vertex.

The main standard algorithm used in CMS for track reconstruction is the Combinatorial Track Finder (CFT) algorithm [7] which uses the reconstructed positions of the passage of charged particles in the silicon detectors to determine the track parameters. The CFT algorithm proceeds in three stages: track seeding, track finding and track fitting. Track candidates are best seeded from hits in the pixel detector because of the low occupancy, the high efficiency and the unambiguous two-dimensional position information.

The track finding stage is based on a standard Kalman filter pattern recognition approach [8] which starts with the seed parameters. The trajectory is extrapolated to the next tracker layer and compatible hits are assigned to the track on the basis of the χ^2 between the predicted and measured positions. At each stage the Kalman filter updates the track parameters with the new hits. In order to take into account possible inefficiencies one further candidate is created without including any hit information. The tracks are assigned a quality based on the χ^2 and the number of missing hits and only the best quality tracks are kept for further propagation. Ambiguities between tracks are resolved during and after track finding. In case two tracks share more than 50% of their hits, the lower quality track is discarded.

For each trajectory the finding stage results in an estimate of the track parameters. However, since the full information is only available at the last hit and constraints

2.2 The CMS Detector

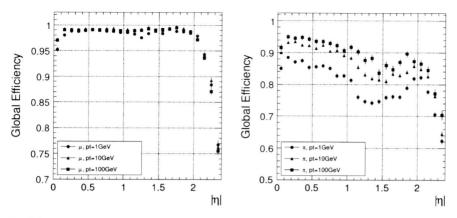

Fig. 2.4 Track reconstruction efficiency for muons (*left*) and pions (*right*) with transverse momenta of 1, 10 and 100 GeV [4]

applied during trajectory building can bias the estimate of the track parameters, all valid tracks are refitted with a standard Kalman filter and a second filter (smoother) running from the exterior towards the beam line.

The expected performance of the track reconstruction is shown in Fig. 2.4 for muons, pions and hadrons. The track reconstruction efficiency for high energy muons is about 99% and drops at $|\eta| > 2.1$ due to the reduced coverage of the forward pixel detector. For pions and hadrons the efficiency is in general lower because of interactions with the material in the tracker. The material budget of the CMS tracker in units of radiation length[1] is presented in Fig. 2.5.

In Fig. 2.6 the transverse momentum resolution for muon tracks with $p_T = 1$, 10 and 100 GeV is shown. At high momenta the resolution is around 1–2% for $|\eta| < 1.6$. The material of the tracker accounts for 20–30% of the transverse momentum resolution. At lower momenta, the resolution is dominated by multiple scattering and its distribution reflects the amount of material traversed by the track. The resolution of the track impact parameters in the longitudinal and the transverse plane are also shown in Fig. 2.6. At high momentum the transverse impact parameter resolution is fairly constant and is dominated by the hit resolution in the first pixel layer. It is progressively degraded by multiple scattering at lower momenta. The same applies to the longitudinal impact parameter resolution. The improvement of the z_0 resolution up to $|\eta| = 0.5$ is due to the charge sharing effects among neighboring pixels.

2.2.5 Electromagnetic Calorimeter

The CMS electromagnetic calorimeter (ECAL) is a hermetic and homogeneous detector with a large pseudorapidity coverage up to $|\eta| < 3$. The ECAL is divided

---

[1] The radiation length X_0 is a characteristic of a material, related to the energy loss of high energy electrons in the material.

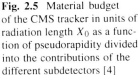

Fig. 2.5 Material budget of the CMS tracker in units of radiation length X_0 as a function of pseudorapidity divided into the contributions of the different subdetectors [4]

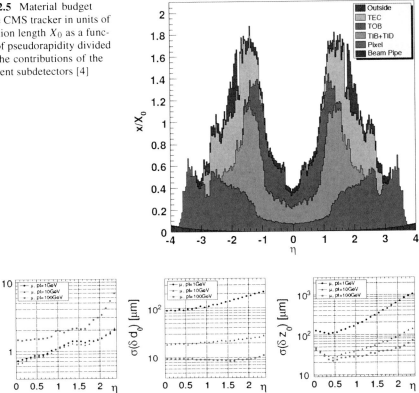

Fig. 2.6 Resolution of track transverse momentum (*left*), transverse impact parameter (*middle*) and longitudinal impact parameter (*right*). The resolution is shown for muons with transverse momenta of 1, 10 and 100 GeV [4]

into barrel and endcap detectors as illustrated in Fig. 2.7. Scintillation crystals made from lead tungstate ($PbWO_4$) are used to measure the energy of electromagnetically interacting particles, mainly electrons and photons. The choice of $PbWO_4$ as the material for the scintillation crystal is motivated by its fast response time and high radiation resistance. The ECAL consists of 61,200 crystals in the barrel and 7,324 crystals in the endcaps. The crystals have a tapered shape and are mounted in a quasi-projective geometry. With a crystal front face of 22×22 mm^2 and 28.6×28.6 mm^2 in the barrel and the endcaps, respectively, a fine granularity is ensured. The length of a barrel crystal is 23 cm which corresponds to 25.8 X_0, while the endcap crystals are 22 cm long corresponding to 24.7 X_0. The scintillation light produced in the crystals is read out by avalanche photo diodes (APD) with an active area of 5×5 mm^2 in the barrel and by vacuum phototriodes (VPT) with an active area of 280 mm^2 in the endcaps. The light output and the amplification have a strong temperature dependence. The response to an incident electron changes by $(3.8 \pm 0.4)\%/°C$ which in

2.2 The CMS Detector

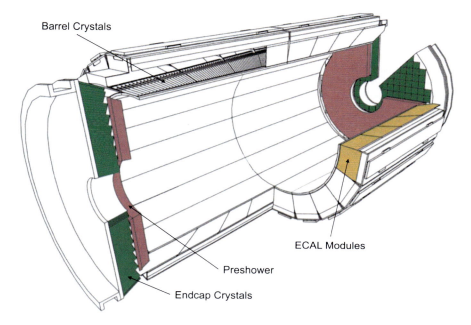

Fig. 2.7 Schematic view of the CMS Electromagnetic Calorimeter [4]

turn means that the temperature has to be closely monitored and kept stable to a precision of ±0.05 °C. The nominal operating temperature of the ECAL is 18 °C and is provided by a water cooling system.

The ECAL barrel has a volume of 8.14 m^3 and its front face is at a radial distance of 1.29 m from the interaction point. It has a 360-fold azimuthal segmentation and two times 85-fold segmentation in pseudorapidity. The endcap has a coverage of $1.479 < |\eta| < 3$ and is situated at a longitudinal distance of 3.15 m from the interaction point. A preshower detector with a thickness of 3 X_0 is placed in front of the endcaps ($1.653 < |\eta| < 2.6$) to guarantee a reliable discrimination of single photons and photons produced in pairs in neutral pion decays.

The energy resolution of the electromagnetic calorimeter can be parametrized by the following expression:

$$\left(\frac{\sigma}{E}\right)^2 = \left(\frac{S}{\sqrt{E}}\right)^2 + \left(\frac{N}{E}\right)^2 + C^2 \qquad (2.1)$$

where S is the stochastic term, N the noise term and C the constant term. The value of the three parameters were determined by a electron test beam measurement to be $S = 0.028\,\text{GeV}^{\frac{1}{2}}$, $N = 0.12\,\text{GeV}$ and $C = 0.003$. The ECAL energy resolution as a function of the energy of electrons is shown in Fig. 2.8.

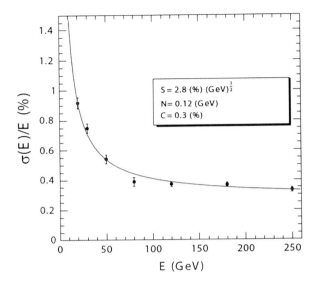

Fig. 2.8 ECAL energy resolution as a function of the energy measured in an electron test beam [4]. The measured values of the stochastic (S), noise (N) and constant (C) term are displayed in the legend

Table 2.1 Tower segmentation in azimuthal and polar angle for the hadronic barrel, endcap and forward calorimeter

	HB/HO	HE $|\eta| \leqslant 2.5$	HE $|\eta| > 2.5$	HF $|\eta| \leqslant 4.7$	HF $|\eta| > 4.7$
$\Delta\phi \times \Delta\eta$	0.087×0.087	0.087×0.087	0.175×0.175	0.175×0.175	0.175×0.35

2.2.6 Hadronic Calorimeter

The energy measurement of the ECAL is complemented by the measurement of the hadronic calorimeter (HCAL). The HCAL is a sampling calorimeter built from alternating layers of massive absorbing brass plates and plastic scintillator tiles arranged in trays. The quality of the energy measurements depends on the fraction of the hadronic shower detected in the calorimeter. Consequently, the thickness of the material in the tray has to be large enough to absorb the major part of the shower. The dimensions of the barrel part of the HCAL however are restricted to the limited volume between the outer extent of the ECAL ($r = 1.77$ m) and the inner extent of the magnetic coil ($r = 2.95$ m). Thus, the HB is supplemented by an outer hadronic calorimeter (HO) located between the solenoid and the muon detectors. The HO uses the solenoid as additional absorbing material and provides sufficient containment for hadronic showers with a thickness of 11.8 interaction lengths (λ_I). The hadronic calorimeter is divided into a barrel part (HB and HO) at $|\eta| < 1.3$, an endcap (HE) on each side at $1.3 < |\eta| < 3$ and a forward calorimeter (HF) extending up to $|\eta| < 5.2$ to achieve a most hermetic detector coverage. The HCAL tower segmentation in the rz plane for one quarter of the HB, HO and HE detectors is shown in Fig. 2.9 and summarized in Table 2.1.

2.2 The CMS Detector

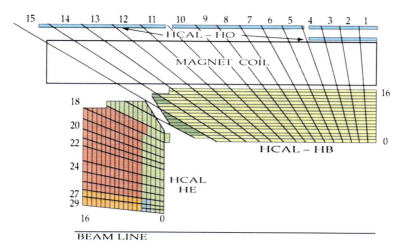

Fig. 2.9 Tower segmentation for one quarter of the HCAL displayed in the rz plane [4]. The colors represent the optical grouping of scintillator layers into different longitudinal readouts

The plastic scintillator tiles are read out by wavelength shifting fibers that shift the blue-violet light emitted by the scintillator to green light which is then sent through transparent fibers to hybrid photodetectors (HPDs) with 19 independent pixels. The first scintillators are placed in front of the first absorber plate in order to sample showers developing in the material between the ECAL and the HCAL, while the last scintillators are installed after the last absorber plate to correct for late developing showers leaking out. 70,000 and 20,916 scintillator tiles are installed in the HB and the HE, respectively.

The HF is positioned at a longitudinal distance of 11.2 m from the interaction point. It will experience unprecedented particle fluxes with an energy of 760 GeV deposited on average in a proton-proton interaction at $\sqrt{s} = 14$ TeV. This energy has to be compared to the average of 100 GeV deposited in the rest of the detector. The situation is even more severe as the energy is not spread equally among the HF, but has a pronounced peak at the highest rapidity. The HF is made from steel absorber plates composed of 5 mm thick grooved plates with quartz fibers inserted as active medium. It detects the Cherenkov light emitted by charged particles in the shower and is thus mainly sensitive to the electromagnetic component of the shower. A longitudinal segmentation in two parts allows to distinguish signals generated by electrons and photons from signals generated by hadrons.

2.2.7 Muon System

The muon system is the outermost part of the CMS detector. The magnet return yoke is equipped with gaseous detector chambers for muon identification and momentum measurement. In the barrel, the muon stations are arranged in five separate iron wheels

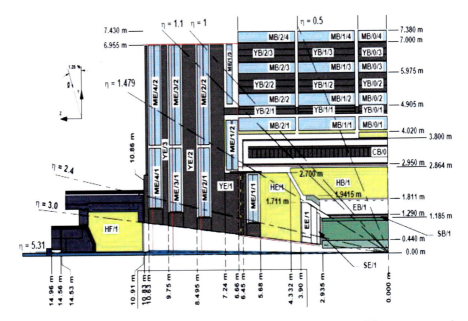

Fig. 2.10 View of one quarter of the CMS detector illustrating the layout of the muon system in the barrel and the endcap region [4]

and in the endcap, four muon stations are mounted onto the three independent iron disks in the positive and the negative endcaps. Each barrel wheel is segmented into 12 sectors in azimuthal angle.

Three different types of gaseous detectors are integrated into the CMS muon system depending on the requirements. In the barrel part where both the muon rate and the neutron induced background are small and the magnetic field is very low, drift tube (DT) chambers are used. In the endcaps however the muon and the background flux is much higher. The muon detector endcaps are thus built from cathode strip chambers (CSCs) which provide a faster response, a higher granularity and a better resistance against radiation. In addition, resistive plate chambers (RPCs) form a redundant trigger system. In total, the CMS muon system consists of 250 DT chambers, 540 CSCs and 610 RPCs. The arrangement of the detector chambers is shown in Fig. 2.10.

A DT cell is a 4 cm wide gas tube with a positively charged stretched wire inside. Each DT chamber, on average 2×2.5 m in size, consists of 12 layers of DT cells, arranged in three groups of four. The middle group measures the z coordinate while the two outside groups measure the $r\phi$ coordinate. In the barrel, four DT chambers are interspersed with the layers of the flux return yoke in each ϕ sector. The outermost muon station is equipped with DT chambers that contain only 8 layers of DT cells and determine the muon position in the $r\phi$ plane. The barrel part of the muon system covers the region $|\eta| < 1.2$.

2.2 The CMS Detector 19

The CSCs are trapezoidal shaped multiwire proportional chambers which consist of 6 anode wire planes crossed with 7 copper strips cathode panels in a gas volume. They provide a two-dimensional position measurement, where the r and ϕ coordinates are determined by the copper strips and the anode wires, respectively. The muon detector endcaps consist of 4 CSC stations on each side and identify muons in the pseudorapidity range of $0.9 < |\eta| < 2.4$.

RPCs are made from two high resistive plastic plates with a voltage applied and separated by a gas volume. The signal generated by the muon when passing through the gas volume is detected by readout strips mounted on top of one of the plastic plates. The RPCs used in the muon trigger system are highly segmented and have a fast response with a time resolution of 1 ns. Six layers of RPCs are installed in the barrel muon system, two layers in each of the first two stations and one layer in each of the last two stations. One layer of RPCs is built into each of the first three stations of the endcap.

2.2.8 Muon Reconstruction

Muon reconstruction, after local-pattern recognition is performed in two stages: stand-alone reconstruction based on information from the muon system only and global reconstruction including the hit information of the silicon tracker. Stand-alone reconstruction starts from track segments in the muon chambers and muon trajectories are built from the inside to the outside using the Kalman filter technique. After the trajectory is built, a second Kalman filter, working from outside in, is applied to determine the track parameters. In the end, the track is extrapolated to the nominal interaction point and a vertex-constrained fit of the track parameters is performed.

In the global muon reconstruction the muon trajectories are extended to add hits measured by the tracker. The track parameters of a stand-alone reconstructed muon are compared to the track parameters of the tracker tracks by extrapolating the trajectories to a common plane on the inner surface of the muon detector. If a tracker track is found that is compatible in momentum, position and direction, the hit information of the tracker and the muon system is combined and refitted to form a global muon track. The resulting global tracks are then checked for ambiguity and quality to choose at most one global track per stand-alone muon.

The precision of the momentum measurement in the muon system is essentially determined by the measurement of the bending angle in the transverse plane at the exit of the magnetic coil. This measurement is dominated by multiple scattering in the material before the first muon station up to transverse momentum values of 200 GeV. For low-momentum muons the momentum resolution is improved substantially by including the measurement of the silicon tracker. The analysis presented here investigates muons with a transverse momentum between 6 and 30 GeV. The inclusion of the tracker information by using global muons is thus most valuable. In Fig. 2.11 a comparison of the momentum resolution of the muon system, the tracker system and a combined measurement is given for the barrel and the forward region.

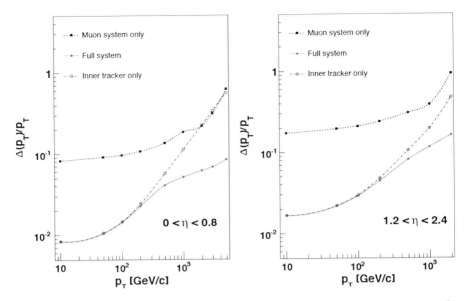

Fig. 2.11 Muon transverse momentum resolution as a function of the transverse momentum for muons detected in the barrel (*left*) and the endcap (*right*) regions [4]. The resolution is given for the measurement using the muon system or the tracking system only and for a combined method

2.2.9 Trigger System

The CMS trigger system is designed to cope with the unprecedented high luminosity and interaction rates. It must ensure high data recording efficiency for a wide variety of physics objects and event topologies, while applying online selective requirements to reduce the 40 MHz event rate to an output rate of about 100 Hz allowing for permanent storage of an event.

The CMS trigger system reduces the event rate in two steps called Level 1 (L1) and High Level Trigger (HLT). The L1 trigger is designed to achieve a maximum output rate of 100 kHz and consists of custom-designed, programmable electronics while the HLT is based on software algorithms running on a large cluster of commercial processors, the event filter farm.

The L1 trigger system uses only coarsely segmented data from the muon system and the calorimeters while the full granularity data are stored in the detector frontend electronics waiting for the L1 decision. The L1 decision has to be taken within a latency time of 3.2 μs and is based on the decision of local, regional and global trigger components. It also depends on the readiness of the other subdetectors and the data acquisition system (DAQ) which is supervised by the Trigger Control System. The trigger architecture is displayed in Fig. 2.12.

The ECAL and the HCAL cells form trigger towers with an (η, ϕ) coverage of 0.087×0.087 in $|\eta| < 1.74$ and larger size in the forward region. Trigger primitives

2.2 The CMS Detector

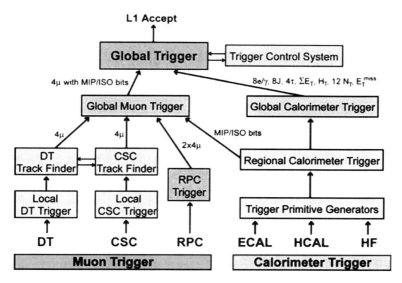

Fig. 2.12 L1 trigger architecture [4]

are generated by calculating the transverse energy of a trigger tower and assigning it to the correct bunch crossing. A regional calorimeter trigger then determines regional electron, photon and jet candidates and information relevant for muon and tau identification. The global calorimeter trigger provides information about the jets, the total transverse energy and the missing energy in the event and identifies the highest-ranking trigger candidates.

In the muon system all three types of detectors take part in the trigger decision. The DT chambers provide track segments in the ϕ projection and hit pattern in η, while the CSC determine three-dimensional track segments. The track finders in the DT chambers and the CSCs calculate the transverse momentum of a track segment and its location and quality. The RPCs deliver an independent measurement derived from regional hit patterns. The global muon trigger receives up to four candidates from each subsystem (DT, barrel RPC, CSC and endcap RPC) together with the isolation information from the global calorimeter trigger. The aim is to improve the efficiency and to reduce the rate by making use of the complementarity and the redundancy of the subsystems. In the end, the global muon trigger selects a maximum of four muon trigger candidates and determines their momentum, charge, position and quality.

The trigger objects extracted by the global calorimeter trigger and the global muon trigger are sent to the global trigger where the decision to accept or reject an event is taken and distributed to the subdetectors. The decision is based on the results of algorithms which for example apply momentum thresholds to single objects or require object multiplicities to exceed predefined values. Up to 128 algorithms can be executed in parallel.

If an event is accepted by the L1 trigger, the full detector information ($\sim$1 MB) is read out by the DAQ system at a rate of up to 100 kHz, passed to the event filter farm and used as input for the HLT. The HLT algorithms are implemented in the same software as used for offline reconstruction and analysis and consist of subsequent reconstruction and selection steps. The framework uses a modular architecture and the modules to be executed are defined at runtime by means of configuration files. The HLT menu is composed of a set of trigger paths, each path addressing a specific physics object selection. The execution of a path is interrupted if the processed event does not fulfill the conditions imposed by a given filter module. The trigger menu for the first CMS data taking can be found in [9].

References

1. O.E. Bruning et al., LHC Design Report, vol I: The LHC Main Ring, CERN-2004-003-V-1 (2004)
2. ATLAS Collaboration, Detector and Physics Performance Technical Design Report, vol. 1, CERN-LHCC-99-14 (1999)
3. CMS Collaboration, The Compact Muon Solenoid: Technical Proposal, CERN-LHCC-94-38 (1994)
4. CMS Collaboration, The CMS experiment at the CERN LHC, JINST **3**, S08004 (2008)
5. LHCb Collaboration, LHCb Technical Proposal, CERN-LHCC-98-04 (1998)
6. ALICE Collaboration, Technical Proposal for a Large Ion Collider Experiment at the CERN LHC, CERN-LHCC-95-71 (1995)
7. W. Adam et al., Track Reconstruction in the CMS Tracker, CMS NOTE-2006/041 (2006)
8. P. Billoir, Comput. Phys. Commun. **57**, 390 (1989)
9. CMS Collaboration, CMS High Level Trigger, CERN/LHCC 2007-021 (2007)

Part I
Study of the Inclusive b Quark Production at CMS

Chapter 3
Heavy Flavor Physics

The study of heavy quark production is an important research area at the LHC. Heavy quarks will be produced with a large cross section at a yet unreached center-of-mass energy, enabling precision measurements to improve our understanding of heavy flavor physics. In the context of this work the term heavy quark stands for charm and beauty quarks since the mass of the up, down and strange quark are significantly lower. The heavier top quark has a very short lifetime and does therefore not form bound states of heavy hadrons.

Heavy quark production is interesting on its own as it presents a key process for the study of the theory of strong interactions, Quantum Chromodynamics (QCD). Furthermore, a well-established theory of heavy quark production is needed for many searches at the LHC.

In this chapter the theoretical concepts relevant to describe the physics of heavy quarks at the LHC are introduced. The main ideas of Quantum Chromodynamics are reviewed, before their application to high-energy hadron-hadron collisions is discussed. This includes the factorization ansatz, the evolution of the parton distribution functions, the partonic processes important for beauty quark production and the phenomenological treatment of heavy quark fragmentation. A further section is dedicated to the description of the decay of b-hadrons via the weak interaction. The Monte Carlo event generators which are used in this analysis to generate full hadronic events within the QCD framework are presented in the last section.

3.1 Quantum Chromodynamics

Quantum Chromodynamics [1–4] is the field theory describing the strong interaction between color charged partons. Color charge comes in three versions (red, green and blue) which form a fundamental representation of the $SU(3)$ symmetry group and is carried by massive spin-$\frac{1}{2}$ quarks and massless spin-1 gluons. Analogous to the photons in Quantum Electrodynamics (QED) [5, 6], the gluons are the gauge bosons

L. Caminada, *Study of the Inclusive Beauty Production at CMS and Construction and Commissioning of the CMS Pixel Barrel Detector*, Springer Theses, DOI: 10.1007/978-3-642-24562-6_3, © Springer-Verlag Berlin Heidelberg 2012

Fig. 3.1 Graphs which contribute to the β function in the one loop approximation: (**a**) fermion loop, (**b**) gluon loop

in QCD and mediate the strong interaction. Since the gluons themselves carry color charge, they can directly interact with other gluons. This possibility is not available in QED, as photons do not have an electric charge. The existence of direct coupling in QCD has important implications on the scale dependence of the strong coupling. A comprehensive overview of QCD applied to hadronic collisions is given in [7, 8].

In QCD, as in any renormalizable quantum field theory, ultraviolet divergences appearing in the calculation can be removed by introducing a scale dependent coupling $\alpha_s(Q^2)$ and a new scale, the renormalization scale μ_R [9]. The dependence of α_s on the energy-scale Q^2 is known as running of the coupling. The requirement that every physical observable has to be independent of the arbitrary choice of μ_R leads to the renormalization group equation

$$Q^2 \frac{\partial \alpha_s(Q^2)}{\partial Q^2} = \beta(\alpha_s). \tag{3.1}$$

While the running of the coupling is determined by the β function in the above equation, the absolute value of α_s has to be determined experimentally. In perturbative QCD (pQCD), the β function can be written as a perturbation series with α_s as expansion parameter. The coefficients of the expansion series are extracted from higher order diagrams, such as the ones shown in Fig. 3.1. Due to the diagrams representing the color self-coupling of the gluons the leading coefficients in QED and QCD have opposite signs. This sign is the origin of the most fundamental difference between QCD and QED. The leading order approximation gives the solution

$$\alpha_s(Q^2) = \frac{1}{b \ln(Q^2/\Lambda_{QCD}^2)}, \quad b = \frac{33 - 2n_f}{12\pi}, \tag{3.2}$$

where n_f is the number of quark flavors with mass below Q^2, and Λ_{QCD} represents the scale at which the perturbative approach is not valid anymore since the coupling becomes too large. Equation (3.2) indicates that in QCD—unlike in QED—the coupling α_s decreases as the energy scale Q^2 increases. This phenomenon is known as asymptotic freedom and justifies the perturbative approach at energy scales $Q^2 > \Lambda_{QCD}$. Experimentally the value of Λ_{QCD} has been found to be of the order of 200 MeV. A summary of measurements of α_s as a function of the respective energy scale Q is presented in Fig. 3.2. At the scale set by the mass of the Z boson the average value of the strong coupling constant is $\alpha_s(M_Z) = 0.1184 \pm 0.0007$ [10].

At energy scales below Λ_{QCD} the strong coupling rises to infinity, which becomes manifest in the confinement of quarks and gluons inside color-singlet hadrons. Quarks

3.1 Quantum Chromodynamics

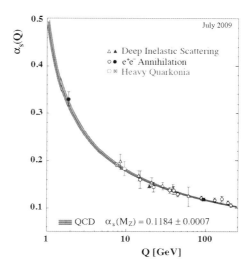

Fig. 3.2 Summary of measurements of α_s as a function of the respective energy scale [10]

and gluons are not seen in experiments, instead they are transformed into observable hadrons in a process called fragmentation. In this regime a perturbative approach is not useful anymore.

Nonetheless perturbative calculations prove successful in the prediction of hadronic cross sections at sufficiently large energy scales. The reason is that the hadronization occurs at a significantly later time scale ($t \sim \frac{1}{\Lambda_{QCD}}$) than the production process ($t \sim \frac{1}{Q}$) and therefore cannot influence the probability of the process to happen.

3.2 Hadronic Collisions

Due to the asymptotic freedom in QCD, the interaction between quarks and gluons becomes arbitrarily weak at short distances. Consequently hadrons behave as collections of free partons at large transferred momenta and their interaction can therefore be described using a parton model.

3.2.1 Event Kinematics

A generic scattering process of two hadrons (h_1, h_2) with four-momenta P_1 and P_2, respectively, is illustrated in Fig. 3.3. The scattering process is caused by the interaction of two partons of the initial hadrons with four-momentum $p_1 = x_1 P_1$ and $p_2 = x_2 P_2$. Since the center-of-mass of the partonic interaction is normally boosted

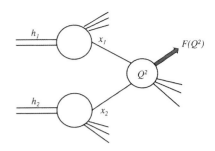

Fig. 3.3 Scattering process of two hadrons h_1 and h_2 in the parton model. Two partons with momentum fractions x_1 and x_2 undergo a hard interaction at the scale Q^2

with respect to the laboratory frame, it is useful to classify the final state according to variables that are invariant under longitudinal boosts. The squared center-of-mass energy of the hadronic system is

$$s = (P_1 + P_2)^2. \tag{3.3}$$

In the massless limit, the virtuality of the process is defined as

$$Q^2 = \hat{s} = x_1 x_2 s. \tag{3.4}$$

The momentum imbalance of the partons participating in the hard interaction is reflected in the rapidity distribution of the outgoing particles. The transverse momentum of the outgoing partons in the center-of-mass frame of the colliding partons is denoted by $\hat{p}_T$ and is of particular interest for the Monte Carlo event generators (see Sect. 3.6).

3.2.2 Factorization

Soft processes resulting in the production of low momentum hadrons will be the most common events in proton-proton collision at the LHC. Although these processes are QCD related, they cannot be calculated by pQCD. Perturbative approaches only lead to reliable results if a hard scale is present in the interaction. In the case of heavy flavor physics, the hard scale is provided by the mass of the heavy quark, its transverse momentum or the virtuality of the process.

Most of the processes calculated by pQCD feature infrared divergences emerging from real gluon emission. Singularities arise either if a gluon is emitted in the direction of the outgoing parton (collinear divergences) or if a low momentum gluon is emitted (soft divergence). Similar to the ultraviolet divergencies which are removed by introducing a renormalization scale, μ_R, the infrared divergencies can be absorbed when imposing a factorization scale, μ_F. The factorization scale can be thought of as the scale which separates the short- and the long-distance physics. The short-

3.2 Hadronic Collisions

distance part covers the hard process calculable in pQCD, while the long-distance part includes the collinear and soft divergencies which are not accessible to perturbative calculations. The factorization ansatz is validated by the factorization theorem [11–13].

According to the factorization theorem the cross section for a hard scattering originating from an interaction of two hadrons with four-momenta P_1 and P_2 can be written as

$$\sigma(P_1, P_2) = \sum_{i,j} \int dx_1 dx_2 f_i^{h_1}(x_1, \mu_F^2) f_j^{h_2}(x_2, \mu_F^2) \hat{\sigma}_{i,j}$$
$$\times (x_1 P_1, x_2 P_2, \alpha_S(\mu_R), Q^2; \mu_F^2, \mu_R^2), \qquad (3.5)$$

where

$f_i^h(x, \mu_F^2)$ is the parton distribution function (PDF) for the parton i in the hadron h,

x_1 is the momentum fraction of the hadron h_1 carried by the parton i,

x_2 is the momentum fraction of the hadron h_2 carried by the parton j,

$\hat{\sigma}_{i,j}$ is the short-distance scattering cross section of partons i and j,

μ_R is the renormalization scale, and

μ_F is the factorization scale.

The parton distribution functions $f_i^h(x, \mu_F^2)$ describe the probability of extracting a parton i from a hadron h with momentum fraction x. Hence, the factorization theorem implies that the probability of extracting the parton can be treated independently from the parton undergoing an interaction. This assumption was successfully verified in deep inelastic lepton-hadron scattering (DIS) which is characterized by a large virtuality ($Q^2 \gg \Lambda_{QCD}$). In the DIS regime, the factorization theorem is proven to be valid to all orders in perturbation theory [13]. Nonetheless it is not obvious that the factorization theorem can be adapted to hadron-hadron collisions since gluons from the hadron remnant might interact and spoil the factorization. Explicit calculations have shown that factorization breaking effects are present but are suppressed by powers of Λ_{QCD} in the high energy limit [14].

The partonic short-distance cross section $\hat{\sigma}_{i,j}$ can be computed in pQCD as

$$\hat{\sigma}_{i,j} = \alpha_s^k \sum_n \left(\frac{\alpha_s}{\pi}\right)^n c^{(n)}. \qquad (3.6)$$

Here the coefficients $c^{(n)}$ are functions of the kinematic variables and the factorization scale. Different hard processes will contribute with different leading powers k to the partonic cross section.

Fig. 3.4 Feynman diagrams contributing to the leading order splitting functions

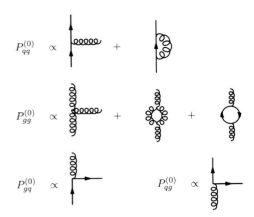

3.2.3 Evolution of Parton Distribution Functions

As discussed in the previous section the long-distance, non-perturbative part of the cross section is absorbed in scale dependent PDFs which cannot be calculated by pQCD. Nevertheless, the dependence on the factorization scale μ_F is described by perturbative calculation. The pQCD parton evolution equations predict the evolution of the PDFs to any scale $Q^2 > Q_0^2$ once $f_i^h(x, Q_0^2)$ is known at a starting scale Q_0^2. The scale dependence of the PDFs is a consequence of gluon radiation and gluon splitting effects, which are incorporated in the DGLAP evolution equations [15–18] for the quark ($q_i(x, Q^2)$) and gluon ($g(x, Q^2)$) PDFs:

$$\frac{dq_i(x, Q^2)}{d \log Q^2} = \int_x^1 \frac{dy}{y} \left(q_i(y, Q^2) P_{qq}\left(\alpha_s(Q^2), \frac{x}{y}\right) + g(y, Q^2) P_{qg}\left(\alpha_s(Q^2), \frac{x}{y}\right) \right) \tag{3.7}$$

$$\frac{dg(x, Q^2)}{d \log Q^2} = \int_x^1 \frac{dy}{y} \left(\sum_i q_i(y, Q^2) P_{gq}\left(\alpha_s(Q^2), \frac{x}{y}\right) + g(y, Q^2) P_{gg}\left(\alpha_s(Q^2), \frac{x}{y}\right) \right) \tag{3.8}$$

where the sum $i = 1, \ldots, 2n_f$ runs over quarks and antiquarks of all flavors. The functions $P_{ab}\left(\alpha_s(Q^2), z\right)$ are called splitting functions and represent the probability to find a parton a in a parton b at the scale Q^2 with a momentum fraction z. The splitting functions are calculable using a perturbative expansion in α_s. The diagrams contributing to the leading order splitting functions are shown in Fig. 3.4.

3.2 Hadronic Collisions

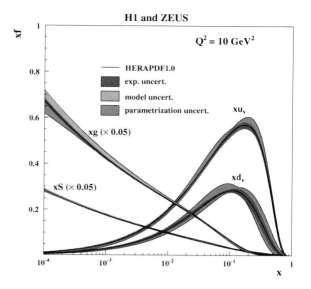

Fig. 3.5 The proton parton distribution functions measured at HERA at $Q^2 = 10\,\text{GeV}$, for valence quarks xu_v and xd_v, sea quarks xS, and gluons xg. The gluon and sea distributions are scaled down by a factor 20 [23]

In practice the PDFs used for calculations in the LHC energy regime are obtained by evolving the PDFs measured in fixed target experiments and in electron-proton scattering at HERA. The standard procedure is to first parametrize the x dependence of the PDFs at a fixed input scale Q_0^2 and then extrapolate the function to the desired scale Q^2 according to the DGLAP equations. Several groups have performed PDF fits to the data obtained in DIS experiments, for example the CTEQ [19], MRST [20], MSTW [21] and NNPDF [22] groups. The quark and gluon distribution functions measured at HERA at $Q^2 = 10\,\text{GeV}$ are shown in Fig. 3.5.

3.3 Heavy Quark Production

The leading-order (LO) process for the production of a heavy quark Q with mass m_Q in hadronic collisions is flavor creation, i.e. quark-antiquark annihilation and gluon-gluon fusion

$$q\bar{q} \to Q\overline{Q} \text{ and } gg \to Q\overline{Q}. \tag{3.9}$$

The corresponding diagrams are shown in Fig. 3.6. When evaluating these diagrams and integrating over the two-body phase space the total partonic cross section at LO in perturbation theory can be obtained [24]. The large energy limit of the partonic cross section is

$$\hat{\sigma}\left(q\bar{q} \to Q\overline{Q}\right) \to \frac{1}{\hat{s}} \tag{3.10}$$

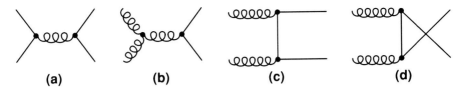

Fig. 3.6 Leading order diagrams for heavy-quark pair production: (**a**) quark-antiquark annihilation $q\bar{q} \to Q\bar{Q}$, (**b**)–(**d**) gluon-gluon fusion $gg \to Q\bar{Q}$

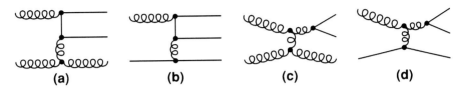

Fig. 3.7 Next-to-leading order diagrams for heavy-quark pair production: (**a**), (**b**) flavor excitation, (**c**), (**d**) gluon splitting

$$\hat{\sigma}\left(gg \to Q\bar{Q}\right) \to \frac{1}{\hat{s}}\left(\frac{1}{\beta}\log\left(\frac{1+\beta}{1-\beta}\right) - 2\right) \tag{3.11}$$

where $\hat{s}$ is the center-of-mass energy available in the partonic system and $\beta \equiv \sqrt{1 - \frac{4m_Q^2}{\hat{s}}}$ is the velocity of the heavy quark. The quark annihilation process vanishes more quickly at high $\hat{s}$ thus gluon-gluon fusion is the dominant process for heavy quark production at the LHC. In flavor creation processes, the final states involving the heavy quarks are observed back-to-back with little combined transverse momentum.

At next-to-leading order (NLO), contributions of real and virtual emission diagrams have to be taken into account. In addition, heavy quarks can be produced in flavor excitation processes and gluon splitting events (Fig. 3.7). In the flavor excitation process, the heavy quark is considered to be already present in the incoming hadron. It is excited by the exchange of a gluon with the other hadron and appears on mass-shell in the final state. Since the heavy quark is not a valence quark it must originate from a pair production process $g \to Q\bar{Q}$. In most PDF parametrizations the heavy-flavor contributions are assumed to vanish for $Q^2 < m_Q^2$, the hard scattering in flavor excitation processes must therefore have a virtuality above m_Q^2. The heavy quark final states do not need to be back-to-back as the third parton can carry away some transverse momentum.

In gluon splitting events the heavy quark occurs in $g \to Q\bar{Q}$ events in the initial- or final-state shower. The resulting heavy flavored final state can carry a large combined transverse momentum and thus be concentrated within a small cone of angular separation. The contribution of the different processes to the total b-quark production cross section predicted by PYTHIA (see Sect. 3.6) is shown in Fig. 3.8 as a function of the center of mass energy.

Fig. 3.8 Total b cross-section as a function of the center-of-mass energy $\sqrt{s}$ in proton-proton collisions. The contributions from pair production, flavor excitation and gluon splitting are shown [25]

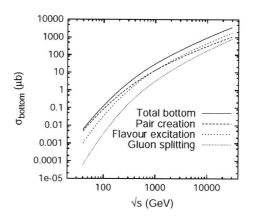

3.4 The Fragmentation of Heavy Quarks

The heavy quarks produced in the hard interaction are not visible in the detector due to color confinement. Instead the quarks fragment into color-singlet hadrons which then conglomerate in a collimated particle jet. As discussed in Sect. 3.1, the fragmentation process happens on a larger time scale compared to the hard process and thus can be treated independently. Since the value of the coupling α_s rises strongly at large distances, the fragmentation process cannot be calculated from first principles in pQCD and phenomenological models have to be applied.

The probability to produce a hadron h from a heavy quark Q can be split in a short- and a long-range part [26]:

$$\mathcal{D}_Q^h(z, \mu_F) = \int_z^1 D_Q(x, \mu_F) \mathcal{D}_Q^h\left(\frac{z}{x}\right) dx. \tag{3.12}$$

The short-distance, perturbative part $D_Q(x, \mu_F)$ models the evolution of a quark produced off-shell at the scale μ_F via gluon emissions to a quark on its mass shell. This is what is usually implemented in the parton shower algorithms of the Monte Carlo simulation programs. A parton shower develops through successive splitting until the perturbative approach becomes unreliable ($\sim \Lambda_{QCD}$). The parton shower represents an approximative perturbative treatment of QCD dynamics based on the DGLAP evolution equations. It improves the fixed order pQCD calculation by taking into account soft and collinear enhanced terms to all orders.

The set of partons in the low-momentum-transfer, long-distance regime produced in the parton shower is transformed into hadrons with the aid of phenomenological fragmentation functions $\mathcal{D}_Q^h(z)$. At present different models of the hadronization process exist among which the string and cluster fragmentation models are of interest for the analysis presented here. Under the assumption of factorization, the fragmentation functions do not depend on the hard scattering process. Hence, the fragmentation

Table 3.1 Peterson parameter for charm and beauty quarks extracted from electron-positron collision data at a center of mass energy $\sqrt{s}$

	$\sqrt{s}$ (GeV)	LO (α_s)	NLO (α_s)
ϵ_c	10.5	0.058	0.035
ϵ_c	91.2	0.078	0.040
ϵ_b	91.2	0.0069	0.0033

The parameters have been determined at LO and NLO [26]

functions are universal and the models tuned for e^+e^- and ep collision data can also be applied to the LHC data.

The string fragmentation model assumes linear confinement. In a physical picture, a color flux tube stretches between the $q\bar{q}$ pair as it starts to move apart. The potential energy stored in the string increases and the string may break by the production of a new quark pair $q_1\bar{q}_1$, so that the system splits into two color-singlet systems $q\bar{q}_1$ and $q_1\bar{q}$. If the invariant mass of either of these string pieces is large enough, further breaking might occur. Gluons are supposed to produce kinks on the strings which modifies the angular distribution of the hadrons inside the jet. The fragmentation function $D_Q^h(z)$ determines the energy and the longitudinal momentum of the hadrons.

The most widely used formula for modeling the fragmentation of heavy quarks is the Peterson fragmentation function. The probability that the hadron receives a momentum fraction z from the quark is given by [27]

$$D_Q^h(z) \propto \frac{1}{z\left(1 - \frac{1}{z} - \frac{\epsilon_Q}{1-z}\right)^2}, \tag{3.13}$$

where ϵ_Q is a free parameter that has to be measured in experiments. It is expected to scale between flavors like $\epsilon_Q \propto 1/m_Q^2$. The values of the Peterson parameter for charm and beauty quarks extracted from electron-positron collision data are listed in Table 3.1. The harder fragmentation of b quarks is explained by their large mass. When binding a light quark to the heavy b quark, the resulting hadron decelerates only slightly so that the b quark and the hadron have almost the same momentum.

In case of cluster fragmentation models color-singlet clusters of partons form after the perturbative phase of jet development and then decay into the observed hadrons. The clusters originate from gluon splitting in quark pairs and subsequent recombination with neighboring quarks and antiquarks. Afterward, the clusters are assumed to decay isotropically in their rest frame into pairs of hadrons, where the branching ratios are determined by the density of states.

3.5 Semileptonic Decays of Heavy Quarks

The presence of hadrons containing heavy quarks is deduced by the observation of their decay products. In a first approximation of b-flavored hadron decays, only the beauty quark participates in the transition while the other quark acts as a specta-

3.5 Semileptonic Decays of Heavy Quarks

Table 3.2 Properties of b-hadrons

	Quark Content	Mass m (MeV)	Lifetime τ (ps)	Decay length $c\tau$ (μm)
B^+	$u\bar{b}$	5279.17 ± 0.29	1.638 ± 0.011	491.1
B^0	$d\bar{b}$	5279.50 ± 0.30	1.525 ± 0.009	458.7
B^0_s	$s\bar{b}$	5366.3 ± 0.6	1.425 ± 0.041	441
Λ^0_b	ubd	5620.2 ± 1.6	$1.383^{+0.049}_{-0.048}$	415

The table shows the quark content, the mass, the lifetime and the decay length [30]

tor quark. The b quark can decay via the weak interaction into a c- or a u-quark. The charged current couplings for the flavor-changing transition between quarks are described in terms of the Cabbibo-Kobayashi-Maskawa (CKM) [28, 29] matrix given by

$$V_{\text{CKM}} = \begin{pmatrix} V_{ud} & V_{us} & V_{ub} \\ V_{cd} & V_{cs} & V_{cb} \\ V_{td} & V_{ts} & V_{tb} \end{pmatrix}. \tag{3.14}$$

The universality of the weak decay is reflected in the unitarity of the CKM matrix. Hence, the CKM matrix can be parametrized by three mixing angles and one irreducible phase which accounts for the CP-violation intrinsic to the weak decay in the Standard Model. The decay width is proportional to the squared CKM matrix element. Measurements of semileptonic decays of B mesons have shown that the matrix elements relevant for the weak decay of the b quark are very small compared to other elements: $|V_{cb}| = 0.0412 \pm 0.0011$ and $|V_{ub}| = 0.00393 \pm 0.00036$ [30]. Consequently, the b quark decay is highly suppressed and the b quark has a relatively large lifetime of $\tau \sim 10^{-12}$ s. Since $|V_{cb}|$ is about an order of magnitude larger than $|V_{ub}|$ the preferred decay is $b \to cW^-$ with a branching ratio of almost 100%.

The lifetime τ of a b-hadron is related to the decay length l by

$$l = \beta\gamma c\tau = \frac{p_B}{m_B} c\tau, \tag{3.15}$$

where p_B, m_B and $\beta\gamma$ are the particle's momentum, mass and boost, respectively. The mean decay length of beauty hadrons is $c\tau = 466\,\mu$m (see also Table 3.2). This transforms into an average observable decay length of $\beta\gamma c\tau = 3$–5 mm in the rest frame at the LHC which can be observed as a displaced (or secondary) vertex in the detector. Objects originating from a secondary vertex are generally characterized by a large transverse impact parameter (Fig. 3.9) and can thereby be identified. In CMS a lifetime based tag of b-hadrons is possible thanks to the pixel detector which achieves a track impact parameter resolution of $\sigma = 80\,\mu$m for $p_T > 7$ GeV and $\sigma = 90\,\mu$m for $p_T = 4$ GeV [31].

The W boson originating from the weak decay of the b quark decays either hadronically or leptonically. Within this analysis the semileptonic decay of b quarks into muons is studied since the muon provides a clean signature which is relatively easy

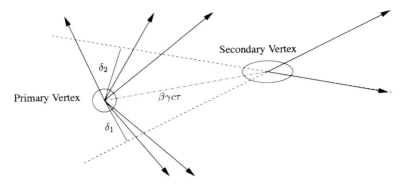

Fig. 3.9 Illustration of the transverse impact parameter (δ_1, δ_2) of the decay products of a long-lived particle. The decay particles emerging from the secondary vertex are characterized by a large impact parameter compared to the impact parameters of the particles emerging from the primary vertex (figure from [32])

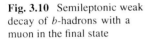

Fig. 3.10 Semileptonic weak decay of b-hadrons with a muon in the final state

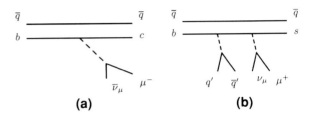

to detect experimentally. The decay $W^- \to \mu^- \bar{\nu}_\mu$ has a branching ratio of about 10%. In addition, about 10% of the subsequent charm decays also have a muon and a neutrino in the final state. The Feynman diagrams of the semileptonic decay of a b-hadron with a muon in the final state are illustrated in Fig. 3.10.

3.6 Monte Carlo Event Generators

Monte Carlo (MC) event generators provide an event-by-event prediction of complete had-ronic final states based on QCD calculation. They allow to study the topology of events generated in hadronic interactions and are used as input for detector simulation programs to investigate detector effects. The event simulation is divided into different stages as illustrated in Fig. 3.11. First, the partonic cross section is evaluated by calculating the matrix element in fixed order pQCD. The event generators presently available for the simulation of proton-proton collisions provide perturbative calculations for beauty production up to NLO. Higher order corrections due to initial and final state radiation are approximated by running a parton shower algorithm. The parton shower generates a set of secondary partons originating from subsequent gluon emission of the initial partons. It is followed by the hadronization algorithm

3.6 Monte Carlo Event Generators

which clusters the individual partons into color-singlet hadrons. In a final step, the short lived hadrons are decayed.

In the framework of the analysis presented here, the MC event generators PYTHIA 6.4 [33–35], HERWIG 6.5.10 [36–38], and MC@NLO 3.4 [39, 40] are used to compute efficiencies, kinematic distributions, and for comparisons with the experimental results. All programs were run with their default parameter settings, except when mentioned otherwise.

PYTHIA

In the PYTHIA program, the matrix elements are calculated in LO pQCD and convoluted with the proton PDF, chosen herein to be CTEQ6L1. The mass of the b-quark is set to $m_b = 4.8\,\text{GeV}$. The underlying event is simulated with the D6T tune [41]. Pile-up events were not included in the simulation. The parton shower algorithm is based on a leading-logarithmic approximation for QCD radiation and a string fragmentation model (implemented in JETSET [42, 43]) is applied. The longitudinal fragmentation is described by the Lund symmetric fragmentation function [44] for light quarks and by the Peterson fragmentation function for charm and beauty quarks. The parameters of the Peterson fragmentation function are set to $\epsilon_c = 0.05$ and $\epsilon_b = 0.005$. In order to estimate the systematic uncertainty introduced by the choice of the fragmentation function, samples generated with different values of ϵ_b are studied.

The hadronic decay chain used in PYTHIA is also implemented by the JETSET program. For comparison, additional event samples are generated where the EvtGen [45] program is used to decay the b-hadrons. EvtGen is an event generator designed for the simulation of the physics of b-hadron decays, and in particular provides a framework to handle complex sequential decays and CP violating decays.

HERWIG

The HERWIG program incorporates color coherence effects in the final state and initial state parton showers, as well as in heavy quark processes and the hard process generation. HERWIG uses a low-mass cluster hadronization model. Multiparton interactions and the underlying event are simulated by the JIMMY package [46]. The value of the b-quark mass is set to $m_b = 4.8\,\text{GeV}$ and the CTEQ6L1 PDF sets are used.

MC@NLO

The MC@NLO package has a NLO matrix element calculation interfaced to the parton shower algorithms of the HERWIG package. The mass of the b-quark is set to $m_b = 4.75\,\text{GeV}$. The CTEQ6M PDF sets were used to generate the MC@NLO events. The events generated with MC@NLO are studied only at the generator level and are not passed through the detailed detector simulation.

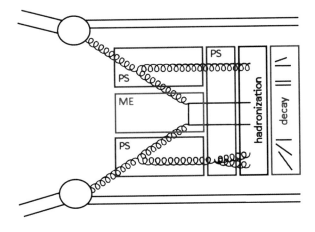

Fig. 3.11 Schematic view of the subsequent steps of a MC event generator: matrix element (ME), parton shower (PS), hadronization and decay

References

1. D.J. Gross, F. Wilczek, Asymptotically free gauge theories. 1. Phys. Rev. D **8**, 3633–3652 (1973)
2. H.D. Politzer, Reliable perturbative results for strong interactions. Phys. Rev. Lett. **30**, 1346–1349 (1973)
3. D.J. Gross, F. Wilczek, Ultraviolet behavior of non-Abelian gauge theories. Phys. Rev. Lett. **30**, 1343–1346 (1973)
4. H.D. Politzer, Asymptotic freedom: an approach to strong interactions. Phys. Rept. **14**, 129–180 (1974)
5. R.P. Feynman, *Quantum Electrodynamics* (W. A. Benjamin, Inc., New York, 1961)
6. S.S. Schweber, *QED and the Men Who Made It: Dyson, Feynman, Schwinger, and Tomonaga* (Princeton University Press, Princeton, USA, 1994), p. 732
7. R.K. Ellis, W.J. Stirling, B.R. Webber, *QCD and Collider Physics*. Cambridge Monographs (Cambridge University Press, Cambridge, 2003)
8. G. Dissertori, I. G. Knowles, M. Schmelling, *Quantum Chromodynamics: High Energy Experiments and Theory* (Oxford Science Publications, Oxford, 2003)
9. G. Sterman et al., Handbook of perturbative QCD. Rev. Mod. Phys. **67**, 157–248 (1995)
10. S. Bethke, The 2009 world average of α_s. Eur. Phys. J. C **64**(2009), 689–703 (2009)
11. R.K. Ellis, H. Georgi, M. Machacek, H.D. Politzer, G.G. Ross, Perturbation theory and the parton model in QCD. Nucl. Phys. B **152**, 285 (1979)
12. J.C. Collins, D.E. Soper, G. Sterman, Factorization of hard processes in QCD. Adv. Ser. Direct. High Energy Phys. **5**, 1–91 (1988)
13. J.C. Collins, D.E. Soper, The theorems of perturbative QCD. Ann. Rev. Nucl. Part. Sci. **37**, 383–409 (1987)
14. J.C. Collins, D.E. Soper, G. Sterman, Nucl. Phys. B **263**, 37 (1986)
15. V.N. Gribov, L.N. Lipatov, e^+e^- pair annihilation and deep inelastic ep scattering in perturbation theory. Sov. J. Nucl. Phys. **15**, 675–684 (1972)
16. V.N. Gribov, L.N. Lipatov, Deep inelastic e p scattering in perturbation theory. Sov. J. Nucl. Phys. **15**, 438–450 (1972)
17. G. Altarelli, G. Parisi, Asymptotic freedom in parton panguage. Nucl. Phys. B **126**, 298 (1977)
18. Y.L. Dokshitzer, Calculation of the structure functions for deep inelastic scattering and e^+e^- annihilation by perturbation theory in quantum chromodynamics. Sov. Phys. JETP **46**, 641–653 (1977)

References

19. J. Pumplin, D.R. Stump, J. Huston, H.L. Lai, P.M. Nadolsky, W.K. Tung, New generation of parton distributions with uncertainties from global QCD analysis. JHEP **0207**, 012 (2002)
20. A.D. Martin, R.G. Roberts, W.J. Stirling, R.S. Thorne, MRST2001: partons and α_s from precise deep inelastic scattering and Tevatron jet data. Eur. Phys. J. C **23**, 73 (2002)
21. A.D. Martin, W.J. Stirling, R.S. Thorne, G. Watt, Parton distributions for the LHC. Eur. Phys. J. C **63**, 189–285 (2009)
22. R. D. Ball et al., [NNPDF Collaboration], A determination of parton distributions with faithful uncertainty estimation. Nucl. Phys. B **809**, 1 (2009) [Erratum-ibid., B **816**, 293 (2009)]
23. H1 and ZEUS Collaborations, Combined measurement and QCD analysis of the inclusive ep scattering cross sections at HERA. JHEP **1001**, 109 (2010)
24. M. L. Mangano, Two lectures on heavy quark production in hadronic collisions. CERN-TH/97-328 (1997)
25. E. Norrbin, T. Sjostrand, Production and hadronization of heavy quarks. Eur. Phys. J. C **17**, 137 (2000)
26. P. Nason, C. Oleari, Nucl. Phys. B **565**, 245 (2000)
27. C. Peterson, D. Schlatter, I. Schmitt, P.M. Zerwas, Scaling violations in inclusive e^+e^- annihilation spectra, Phys. Rev. D **27** (1983) 105
28. N. Cabibbo, Unitarity symmetry and leptonic decays. Phys. Rev. Lett. **10**, 531–533 (1963)
29. M. Kobayashi, T. Maskawa, CP-violation in the renormalizable theory of weak interaction. Prog. Theor. Phys. **49**, 652–657 (1973)
30. C. Amsler et al., Particle data group. Phys. Lett. B **667**, 1 (2008)
31. H.C. Kästli et al., CMS barrel pixel detector overview. Nucl. Instrum. Meth. A **582**, 724 (2007)
32. R. Weber, Diffractive ρ^0 photoproduction at HERA. Ph.D. Thesis, ETH Zürich, ETH No. 16709, 2006
33. H.U. Bengtsson, T. Sjostrand, Comp. Phys. Comm. **46**, 367 (1987)
34. T. Sjostrand, Comp. Phys. Comm. **82**, 74 (1994)
35. T. Sjostrand, S. Mrenna, P.Z. Skands, PYTHIA 6.4 physics and manual. JHEP **05**, 026 (2006)
36. G. Marchesini, B.R. Webber, Nucl. Phys. B **310**, 461 (1988)
37. G. Marchesini, B.R. Webber, G. Abbiendi, I.G. Knowles, M.H. Seymour, L. Stanco, Comp. Phys. Comm. **67**, 465 (1992)
38. G. Corcella et al., HERWIG 6.5: an event generator for hadron emission reactions with interfering gluons (including supersymmetric processes). JHEP **01**, 010 (2001)
39. S. Frixione, B.R. Webber, Matching NLO QCD computations and parton shower simulations. JHEP **0206**, 029 (2002)
40. S. Frixione, P. Nason, B.R. Webber, Matching NLO QCD and parton showers in heavy flavour production. JHEP **0308**, 007 (2003)
41. ATLAS, CMS, TOTEM Collaboration, Multiple parton interactions, underlying event and forward physics at LHC. in *Proceedings of Multiple Parton Interactions at the LHC, 1st Workshop*, DESY-PROC-2009-06 (2008)
42. T. Sjostrand, Comp. Phys. Comm. **39**, 347 (1986)
43. M. Bengtsson, T. Sjostrand, Comp. Phys. Comm. **43**, 367 (1987)
44. B. Andersson, G. Gustafson, B. Söderberg, Z. Phys. C **20**, 317 (1983)
45. http://www.slac.stanford.edu/~lange/EvtGen/
46. J.M. Butterworth, J.R. Forshaw, M.H. Seymour, Multiparton interactions in photoproduction at HERA. Z. Phys. C **72**, 637–646 (1996)

Chapter 4
Study of the Inclusive Beauty Production

The prospects for a measurement of the inclusive b-quark production cross-section with the very early data at CMS are presented in this chapter. Due to the large b-quark production cross-section, high statistics data samples are expected soon after the LHC start-up. Measurements of the b-quark production have already been done at the Tevatron [1–4], HERA [5–8] and other colliders. A lot of progress has been made in understanding the b-quark production process and the measurements are in reasonable agreement with NLO/NLL QCD predictions in most regions of the phase space. However, theoretical uncertainties are sizeable and there is a great interest in verifying the results at the higher center-of-mass energy provided by the LHC. The investigation of b-quark production is interesting on its own, but also because events containing b quarks present an important background to most of the searches at the LHC. In the first section the concept of the measurement is introduced, followed by an overview of the event samples used in the analysis. Thereafter the trigger and offline selection are discussed. The fitting procedure to determine the fraction of signal events among the selected events is described and data-driven methods for validating the MC templates used in the fit are presented. Furthermore, the method for extracting the inclusive and the differential b-quark production cross-section is reviewed. The chapter closes with a discussion of the main systematic uncertainties and the results.

4.1 Concept

In the context of this analysis the semileptonic decay of b quarks into muons and jets is studied. The muons provide a clean signal in the detector which permits to identify them already on trigger level. Muons from b- and c-quark decays can be distinguished by their momentum distribution. Due to the higher mass of the b quark ($m_b = 4.95\,\text{GeV}$) more energy is transferred to the daughter particle on average. Furthermore, the transverse momentum of the muon relative to the fragmentation

L. Caminada, *Study of the Inclusive Beauty Production at CMS and Construction and Commissioning of the CMS Pixel Barrel Detector*, Springer Theses, DOI: 10.1007/978-3-642-24562-6_4, © Springer-Verlag Berlin Heidelberg 2012

Fig. 4.1 Illustration of the $p_\perp^{rel}$ variable which denotes the component of the muon momentum perpendicular to the jet direction

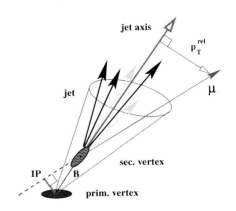

jet ($p_\perp^{rel}$) is larger in b-events than in c- and light quark events. The $p_\perp^{rel}$ variable is illustrated in Fig. 4.1 and will be used in this analysis to discriminate signal events from background events.

At a center-of-mass energy of $\sqrt{s} = 7\,\text{TeV}$, an inclusive b-quark production cross-section of $\sigma_b = 322\,\mu\text{b}$ is predicted by PYTHIA. The simulated transverse momentum and pseudorapidity distributions of the b quark are shown in Fig. 4.2. In about 20% of the cases a muon is found among the decay products of the b quark. In Fig. 4.3 the transverse momentum and the pseudorapidity of the muon generated either in the semileptonic decay of the b quark or the subsequent c quark is displayed. The visible kinematic range for muons in the measurement presented here is $p_T > 5\,\text{GeV}$ and $-2.1 < \eta < 2.1$, which corresponds to an acceptance of 2%.

The $p_\perp^{rel}$ variable is defined with respect to the axis of the fragmentation jet which is reconstructed from charged particle tracks only. The tracks are clustered by a jet algorithm (TrackJets) and a combined transverse energy of $E_T > 1\,\text{GeV}$ is required. In order to facilitate the comparison between the measurement and the theory predictions, the TrackJet is not included in the cross section definition. The extrapolation of the measured cross-section amounts to 10%.

4.2 Event Simulation

The MC event samples were generated, simulated and reconstructed within the official CMS software framework (CMSSW) versions CMSSW_2_2_X and CMSSW_3_1_X. The generation of Monte Carlo event samples is based on PYTHIA V6 and does not include pile-up events. All QCD signal and background events are selected from the generic 2 → 2 subprocesses (default PYTHIA MSEL = 1 card) and present a mixture of gluon-gluon fusion, flavor excitation, and gluon splitting events. No attempt is made to separate the production mechanisms and all

4.2 Event Simulation 43

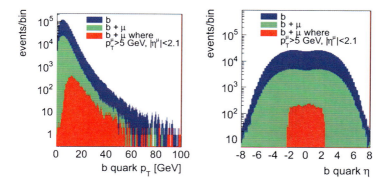

Fig. 4.2 Transverse momentum (*left*) and pseudorapidity (*right*) distribution of *b* quarks originating from proton-proton collisions at a center-of-mass energy of $\sqrt{s} = 7$ TeV. The inclusive distribution is shown in *blue*, the *green* distribution corresponds to *b* quarks that decay semileptonically into muons and the *red* one describes quarks whose decay produce muons within the visible kinematic range ($p_T > 5$ GeV and $-2.1 < \eta < 2.1$)

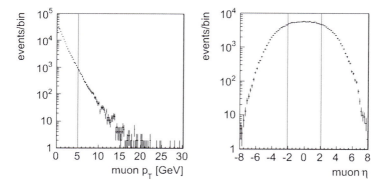

Fig. 4.3 Generated transverse momentum (*left*) and pseudorapidity (*right*) distribution of muons produced in the semileptonic decay of *b* quarks. The *lines* at $p_T = 5$ GeV and $|\eta| = 2.1$ indicate the visible kinematic range

events containing a *b* quark are counted as *b*-events. Events containing a *c* quark but no *b* quark are called *c*-events. All remaining events are called *udsg*-events.

The samples were simulated in separate $\hat{p}_T$ ranges and event weights are introduced to scale the contribution of the individual $\hat{p}_T$ bins according to the cross section predicted by PYTHIA.

A muon filter is applied to all generated events. A special procedure has been used to efficiently simulate inclusive muon samples including prompt muons and muon originating from in-flight decays of pions and kaons. An inclusive muon sample is achieved by forcing kaons and pions to decay already during generation rather than during simulation. The generated muon has to be in the kinematic range of $-2.5 < \eta < 2.5$ and $p_T < 5$ GeV, except for the two QCD samples with $\hat{p}_T > 0$ GeV

Table 4.1 Overview of the event samples used in the analysis

Sample	$\sqrt{s}$ (TeV)	σ (μb)	$\mathcal{L}$(pb^{-1})	ϵ_{filt}	N_{filt}	$N_{\text{filt}}^{b}/N_{\text{filt}}$
Signal						
$\hat{p}_T > 20\,\text{GeV}$	10	497.2	2.9	0.0037	5,292,356	1
QCD (inclusive)						
$\hat{p}_T > 0\,\text{GeV}$	10	51,560	0.04	0.0023	4,844,040	0.20
$\hat{p}_T > 20\,\text{GeV}$	10	497.2	2.4	0.0079	9,623,100	0.47
$30 < \hat{p}_T < 50\,\text{GeV}$	10	91.78	9.2	0.01	8,411,460	0.46
$50 < \hat{p}_T < 80\,\text{GeV}$	10	11.66	10.3	0.021	2,511,780	0.44
$80 < \hat{p}_T < 120\,\text{GeV}$	10	1.55	14.4	0.033	735,244	0.43
$120 < \hat{p}_T < 170\,\text{GeV}$	10	0.25	30.0	0.044	328,590	0.42
$\hat{p}_T > 170\,\text{GeV}$	10	0.06	125.3	0.06	466,279	0.40
$\hat{p}_T > 0\,\text{GeV}$	7	48,440	0.12	0.0018	10,314,205	0.21

In the first three columns, the center-of-mass energy of the proton-proton collision, the cross section (PYTHIA MSEL = 1 card) and the integrated luminosity of the samples are given. The remaining columns state the filter efficiency of the generator level filter, the number of events that pass the filter and the fraction of b-events among the filtered events. The transverse momentum threshold for the generated muon is $p_T > 5\,\text{GeV}$, except for the two QCD samples with $\hat{p}_T > 0\,\text{GeV}$ where it is lowered to $p_T > 2.5\,\text{GeV}$

where it is lowered to $p_T < 2.5\,\text{GeV}$. For signal events, an additional filter is applied requiring a b quark in the generated event.

The events undergo a full detector simulation based on GEANT4 [9] including a emulation of the first level trigger. Subsequent high level triggers and offline reconstruction use the same code as applied to data.

Table 4.1 provides a summary of the event samples used in the analysis. If not stated otherwise the MC simulation corresponding to a center-of-mass energy of $\sqrt{s} = 7\,\text{TeV}$ is used to obtain the results in this chapter.

4.3 Trigger

The events of interest are selected by requiring at least one single muon on trigger level. The muon identification on trigger level is based on subsequent reconstruction and filtering steps. As discussed in Sect. 2.2.9, the decision of the L1 muon trigger is based on the information of the muon system only. The measurement of the muon system and the tracking detectors are combined at the HLT in case a muon candidate is present at L1. First a standalone (L2) muon reconstruction is performed using the parameters of the L1 muon candidate for seeding. If a standalone muon with transverse momentum above threshold is found, it serves in turn as a seed for the global muon (L3) reconstruction. After the reconstruction has terminated, further cuts are applied on the transverse momentum of the global muon and the track impact parameter in the transverse plane ($d_0 < 2\,\text{cm}$). An event is accepted on trigger level provided that a L3 muon passing the last filter module is present.

4.3 Trigger

Table 4.2 Transverse momentum threshold (in GeV) applied to the muon candidates on trigger level for the HLT paths investigated in this analysis

	L1	L2	L3
HLT_Mu3	0	3	3
HLT_Mu5	3	4	5

Table 4.3 Expected HLT transverse momentum threshold for the single muon and di-muon trigger paths for different luminosities

Luminosity $[\mathrm{cm}^{-2}\mathrm{s}^{-1}]$	Single muon	Di-muon
$8 \cdot 10^{29}$	3	3/3
10^{31}	5	3/3
$2 \cdot 10^{31}$	11	3/3
10^{32}	16	3/3

The performance of the muon trigger has been studied in detail in [10]. In the framework of this analysis two HLT paths (HLT_Mu3 and HLT_Mu5) which mainly differ in the transverse momentum threshold ($p_T > 3(5)\,\mathrm{GeV}$) are investigated. The L1 and HLT conditions for these paths are listed in Table 4.2.

The muon transverse momentum threshold on L1 and HLT have been optimized for the LHC startup assuming an instantaneous luminosity of $L = 8 \cdot 10^{29}\mathrm{cm}^{-2}\mathrm{s}^{-1}$. When the luminosity delivered by the LHC increases, the single muon trigger will have a higher rate which at some point will exceed the allocated bandwidth. In this case either the transverse momentum threshold has to be increased or prescale factors have to be applied. Table 4.3 shows the muon trigger transverse momentum threshold values projected for increased values of the LHC luminosity.

The trigger efficiency is determined by a trigger simulation. The combined L1 and HLT efficiency for signal events are displayed in Fig. 4.4 as a function of muon transverse momentum and pseudorapidity. The efficiency is calculated with respect to the number of generated particles in the acceptance of the measurement, i.e $|\eta| < 2.1$ and $p_T > 5\,\mathrm{GeV}$. For HLT_Mu3 and HLT_Mu5 the efficiency plateau is reached for muons with $p_T > 18$–$20\,\mathrm{GeV}$ and lies around 90%. The slightly lower efficiency of HLT_Mu5 is due to the more restrictive L1 condition. Besides the higher transverse momentum also a better quality (see Sect. 2.2.9) of the muon candidate is required.

The overall trigger efficiency for $b \rightarrow \mu X$ events that have a muon with $p_T > 5\,\mathrm{GeV}$ in $-2.1 < \eta < 2.1$ on generator level is 84% for HLT_Mu3, while it is 79% for HLT_Mu5.

4.4 Jet Reconstruction

The use of a collinear and infrared safe jet reconstruction algorithm in an analysis is strongly recommended when comparing the results to the theoretical predictions. In this analysis jets are reconstructed by the anti-k_T algorithm [11]. The algorithm is

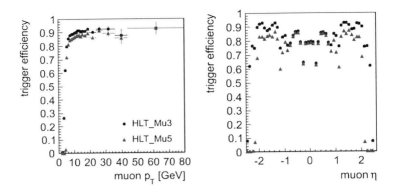

Fig. 4.4 The single muon trigger efficiency as a function of the muon transverse momentum and pseudorapidity in signal events for HLT_Mu3 (*circles*) and HLT_Mu5 (*triangles*). In the *left plots* the efficiencies are determined with respect to generated muons with pseudorapidity $-2.1 < \eta < 2.1$, in the *right plots* for muons with transverse momentum $p_T > 5\,\text{GeV}$

based on successive pair-wise recombination of particles according to the distance between any two particles i and j (d_{ij}) and the distance of any particle i to the beam (d_{iB}) which are defined as

$$d_{ij} = \min\left(k_{Ti}^{-2}, k_{Tj}^{-2}\right) \frac{[(y_i - y_j)^2 + (\phi_i - \phi_j)^2]}{D^2},$$

$$d_{iB} = k_{Ti}^{-2},$$

where k_{Ti}, y_i and ϕ_i are the transverse momentum, the rapidity and the azimuthal angle of particle i, respectively. Jets are defined in an iterative procedure. If the smallest distance is a d_{ij}, the corresponding particles are combined, otherwise particle i is defined as a jet. The anti-k_T algorithm is reasonably fast and collinear and infrared safe to all orders in the perturbative expansion. The jet size parameter D gives the minimum distance between two jets and a value of $D = 0.5$ is chosen here.

Three kinds of jet objects are produced by the SISCone algorithm: calorimeter jets (CaloJets), tracker jets (TrackJets) [12] and generated particle jets (GenJets). The latter are clustered from all stable generated particles and are a measure of the hadron-level jet content. The reconstruction of CaloJets is based on energy depositions in calorimeter towers. A calorimeter tower is built from one HCAL cell and 5×5 ECAL crystals in the barrel, while a more complex association is required in the endcaps [13]. Three stages of jet corrections are applied offline to CaloJets [14]: offset corrections for pile-up and noise, correction of the calorimeter response as a function of pseudorapidity and correction for the absolute response as a function of transverse momentum.

TrackJets are obtained when using only charged particle tracks as input to the jet algorithm. They are expected to give a better result when reconstructing low

4.4 Jet Reconstruction

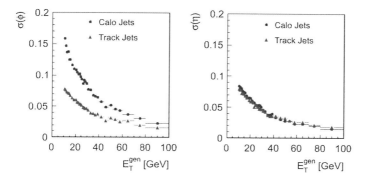

Fig. 4.5 The jet angular resolution as a function of the transverse energy of the generated jet. The *left plot* shows the resolution in azimuthal angle, the *right plot* the resolution in pseudorapidity. The angular resolution of TrackJets (*triangles*) is compared to the resolution of CaloJets (*circles*)

Fig. 4.6 The average difference between the transverse energy of the generated and the reconstructed jet as a function of the transverse energy of the generated jet. The energy response of TrackJets (*triangles*) is compared to the energy response of CaloJets (*circles*). The three levels of jet corrections are applied to the CaloJets. The errors correspond to the width of a Gaussian distribution fitted to the core of the distribution of the residuals

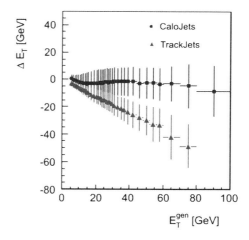

transverse energy jets. This is due to the fact that low momentum charged particles might not reach the calorimeter or might be strongly deflected from the original direction. The use of TrackJets also provides an independent measurement with its own systematic uncertainty. It has the additional benefit of an improved jet angular resolution (mainly in ϕ) as can be seen in Fig. 4.5. Here, the jet resolution in azimuthal angle and pseudorapidity is presented for events in which a b quark decays semileptonically into a muon and a jet. A good understanding of the jet angular resolution and its uncertainty is crucial for the measurement as the relative transverse momentum of the muon with respect to the jet is computed to discriminate signal events from background events. The disadvantage of using jets reconstructed from tracks only is their low energy response (Fig. 4.6). The energy of a TrackJet will always be lower than the energy of the corresponding GenJet since the fraction of jet momenta carried by neutral particles is not visible in the tracking detectors.

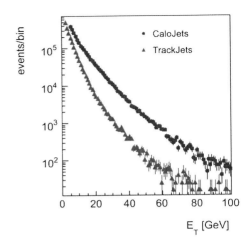

Fig. 4.7 Transverse energy spectrum of b-jets reconstructed from TrackJets and CaloJets. The mean transverse energy is $E_T = 5.7\,\text{GeV}$ for TrackJets and $E_T = 12.2\,\text{GeV}$ for CaloJets

In most cases the muon emitted in the decay of a b quark is reconstructed within the jet. Since the event selection developed for the analysis presented here starts from the muon reconstructed on trigger level, the jet containing the muon is identified as a b-jet. Figure 4.7 shows the transverse energy spectrum for b-jets reconstructed from CaloJets and TrackJets, respectively. The spectrum is falling steeply and a mean value of $E_T = 5.7\,\text{GeV}$ for TrackJets and $E_T = 12.2\,\text{GeV}$ for CaloJets is found. The transverse energy of the TrackJet is calculated by summing the four-momenta of its constituents which are assumed to be pions. The muon is not considered as a jet constituent, i.e. the muon track is excluded from the TrackJet. The three stages of jet corrections are applied to correct the response of the CaloJets back to the particle level jet, whereas the CaloJet is not corrected for the energy deposition of the muon in the calorimeter. The size of the correction depends on the transverse energy of the generated jet. According to [14] correction factors of the order of 2.5 to 3.2 have to be applied for jets with $E_T < 10\,\text{GeV}$. These numbers are currently determined from simulations and affected by a significant systematic uncertainty. In order to avoid entering the phase space region in which large systematic uncertainties on the cross section are introduced due to the large jet energy correction factors, the CaloJets are required to have a minimum raw transverse energy of $E_T > 10\,\text{GeV}$.

TrackJets prove to be more reliable when going to low jet energies. Within this analysis, TrackJets are reconstructed from all charged tracks with transverse momentum $p_T > 0.3\,\text{GeV}$ and a minimum transverse energy of the TrackJet of $E_T > 1\,\text{GeV}$ is required. In addition cuts on the number of pixel and strip detector layers with measurements are applied in order to ensure a good quality of the tracks. A summary of the selection criteria for the tracks used in the clustering algorithm is presented in Table 4.4. The number of tracks reconstructed within a TrackJet is shown in Fig. 4.8.

The efficiency of associating a jet to the muon depends on the muon transverse momentum and is displayed in Fig. 4.9 for signal events. The efficiency is calculated for events in which a muon and a b-jet are present at generator level. Figure 4.9 also

4.4 Jet Reconstruction

Table 4.4 Selection criteria for the tracks used in the TrackJet clustering algorithm

Variable	Value
Minimum transverse momentum	$p_T > 0.3$ GeV
Maximum transverse momentum	$p_T < 500$ GeV
Longitudinal impact parameter	$z_0 < 15$ cm
Number of pixel layers with hits	$\geqslant 2$
Number of tracker layers with hits	$\geqslant 5$

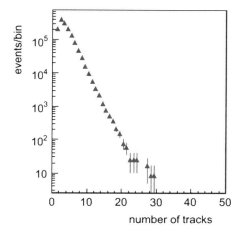

Fig. 4.8 Distribution of the number of tracks reconstructed within a TrackJet

shows the fraction of mistagged b-jets as a function of the jet transverse energy. It is estimated by comparing the reconstructed jets to the generated b-jet. If the generated b-jet lies within a cone of radius $R = \sqrt{\eta^2 + \phi^2} = 0.5$ around the reconstructed jet, the matching is successful and the tagging is defined to be correct. The fraction of mistagged b-jets is obtained by dividing the number of mistagged selected jets by the total number of selected jets in signal events.

The efficiency rises from 68% to almost 100% for events containing a muon with transverse momentum $p_T > 20$ GeV. The fake rate in the lowest bin in muon transverse momentum takes values up to 10% and decreases for higher muon transverse momenta. If the measurement with data showed that a large fake rate makes the control of the background difficult, the lower cut on the transverse energy of TrackJets would have to be increased. The efficiency for CaloJets in the low muon transverse momentum bins is significantly lower compared to TrackJets, while the fake rate is higher.

Within this study the use of TrackJets is proposed for the analysis of the very first data of CMS. At increased LHC luminosity when the focus will be on higher momentum muons the use of CaloJets as suggested in [15, 16] becomes interesting.

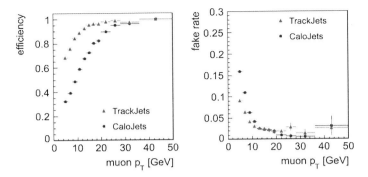

Fig. 4.9 *Left* efficiency for finding a TrackJet (*triangles*) or CaloJet (*circles*) in a cone of radius $R = 0.5$ around the muon as a function of the muon transverse momentum. *Right* fraction of mistagged b-jets as a function of the muon transverse momentum

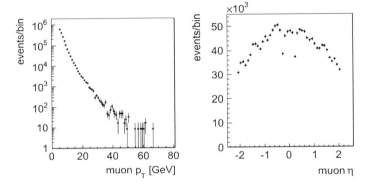

Fig. 4.10 Simulated distribution of reconstructed muon transverse momentum and pseudorapidity in signal events passing the event selection. The distributions are normalized to an integrated luminosity of $1\,\text{pb}^{-1}$

4.5 Event Selection

As discussed in the previous section, the first requirement of the event selection is a reconstructed muon with transverse momentum $p_T > 5\,\text{GeV}$ and pseudorapidity $-2.1 < \eta < 2.1$. Only global muons, i.e. muons identified in the tracking detectors as well as in the muon chambers, pass the selection. If more than one muon is present in the event, the muon with the highest transverse momentum is chosen. The transverse momentum and pseudorapidity distributions of muons passing the event selection are shown in Fig. 4.10 for b-events.

The assignment between the reconstructed muon and the TrackJet is based on the result of the TrackJet clustering algorithm. The muon is associated to a TrackJet if the muon track lies within the TrackJet with transverse energy $E_T > 1\,\text{Gev}$. The

4.5 Event Selection

Fig. 4.11 Distance between the reconstructed muon and the TrackJet (ΔR) for selected b-, c- and $udsg$-events. The event selection requires a muon with transverse momentum $p_T > 5\,\mathrm{GeV}$ and pseudorapidity $|\eta| < 2.1$ and a TrackJet with transverse energy $E_T > 1\,\mathrm{GeV}$. The distributions are normalized to an integrated luminosity of $1\,\mathrm{pb}^{-1}$

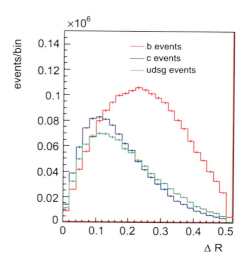

distance between the muon and the TrackJet

$$\Delta R = \sqrt{\Delta\eta^2 + \Delta\phi^2}$$

in signal and background events passing the event selection is shown in Fig. 4.11. A total number of 3.7 million events are selected in a data set corresponding to $1\,\mathrm{pb}^{-1}$ of integrated luminosity. Of those, 49% contain a b quark while 26% are classified as c-events and the remaining 25% as $udsg$-events. The variation of the b fraction with the transverse momentum and the pseudorapidity of the muon is shown in Fig. 4.12 for simulated events.

The relative transverse momentum of the muon with respect to the axis of the associated jet is calculated according to the formula

$$p_\perp^{\mathrm{rel}} = \frac{|\vec{p}_\mu \times \vec{p}_{jet}|}{|\vec{p}_{jet}|},$$

where $\vec{p}_\mu$ and $\vec{p}_{jet}$ are the momentum vector of the muon and the TrackJet, respectively.

The $p_\perp^{\mathrm{rel}}$ distribution for b-, c- and $udsg$-events is shown in Fig. 4.13. The b-spectrum contains a contribution from cascade decays $b \to c + X \to \mu + X'$ in addition to the direct semileptonic decays $b \to \mu + X$. The $p_\perp^{\mathrm{rel}}$ spectrum of b-events is significantly harder than the $p_\perp^{\mathrm{rel}}$ spectrum of c- and $udsg$-events, while the c- and $udsg$-spectrum turn out to be rather similar.

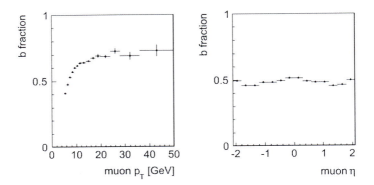

Fig. 4.12 Fraction of b-events in the inclusive sample as a function of the transverse momentum and the pseudorapidity of the reconstructed muon

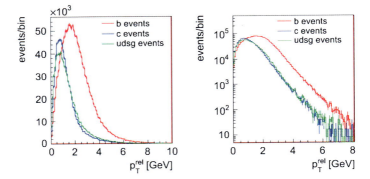

Fig. 4.13 $p_\perp^{rel}$ distribution of the inclusive sample separated by the quark content and normalized to an integrated luminosity of $1\,\text{pb}^{-1}$ in selected events. The event selection requires a muon with transverse momentum $p_T > 5\,\text{GeV}$ and pseudorapidity $|\eta| < 2.1$ and a TrackJet with transverse energy $E_T > 1\,\text{GeV}$. The *left figure* shows the $p_\perp^{rel}$ distribution on a linear scale, while in the *right figure* a logarithmic scale is used

4.6 Signal Extraction

A fit to the observed $p_\perp^{rel}$ spectrum based on simulated templates is used to determine the fraction of signal events among all events passing the event selection. As the $p_\perp^{rel}$ spectrum in c- and $udsg$-events proves to be very similar, it cannot be separated in the fitting procedure. Therefore the c- and $udsg$-background component are combined and a fit with one signal component against one background component is implemented. The uncertainty of the unknown c to $udsg$ background composition is treated as a systematic uncertainty (see Sect. 4.9).

4.6 Signal Extraction

4.6.1 Fitting Procedure

The estimation of the sample composition based on MC simulations of the individual sources is a common problem in experimental data analysis. The contribution of different sources is determined by fitting the binned observed data with template histograms obtained by simulation. A fitting method which takes into account the limited statistics of the MC simulation is detailed in [17] and the main points are reviewed here. The goal is to determine the scale factors α_j for each source j which relate the (unknown) number of expected MC events A_{ji} to the predicted number of data events f_i in each bin i

$$f_i = \sum_{j=1}^{m} \alpha_j A_{ji}, \tag{4.1}$$

where m is the number of sources. Depending on the binning the number of events per bin might be quite small and the probability of observing a particular number of data events d_i in bin i is thus best described by Poisson statistics

$$\mathcal{P}(d_i \mid f_i) = \frac{f_i^{d_i}}{d_i!} e^{-f_i}. \tag{4.2}$$

Analogous to the expected number of data events f_i, the expected number of MC events A_{ji} represent the mean values of the Poisson distributed number of generated MC events a_{ji} of source j in bin i. The introduction of the unknown parameters A_{ji} accounts for the fact that also the number of generated MC events fluctuates because of the limited size of the available MC samples.

Estimates for the parameters α_j and A_{ji} are determined by using the maximum likelihood technique and maximizing the negative logarithm of the likelihood function $\mathcal{L}$

$$- \ln \mathcal{L} = \sum_{i=1}^{n} d_i \ln f_i - f_i + \sum_{i=1}^{n} \sum_{j=1}^{m} a_{ji} \ln A_{ji} - A_{ji}. \tag{4.3}$$

In contrast to the results for the parameters α_j which are the scale factors we are interested in, the parameters A_{ji} are only a mathematical tool and the result is of no importance for the measurement. When differentiating equation (4.3) with respect to α_j and A_{ji} and setting the derivatives to zero a set of $m \times (n+1)$ non-linear and coupled equation is obtained. The equation can be simplified to derive the dependence of A_{ji} on the m variables p_j in each bin separately. Thereafter the solution for p_j can be found iteratively. The ROOT package [18] provides an implementation of the fitting procedure (TFractionFitter [19]) where the optimization problem is solved by MINUIT [20].

Special care has to be taken when weights are imposed on the MC events since the assumption that the number of MC events a_{ji} obey Poisson statistics is not valid anymore. Within this analysis weights are introduced to scale the contributions of the individual $\hat{p}_T$ bins according to their cross section. The event weights are given by

$$w_k = \frac{\sigma_k \cdot \mathcal{L}_{tot}}{N_k}, \tag{4.4}$$

where N_k is the number of generated events in bin k, σ_k is the theoretical cross section of the corresponding simulated process and $\mathcal{L}_{tot}$ the integrated luminosity.

In the following, two procedures developed to accommodate the weights in the fitting procedure are discussed. While the first method represents an approximative treatment which is only valid if the spread of the weights among the bins is not too large, the full treatment is covered by the second method using an extension of the likelihood function.

1. Approximate Solution

The fitting procedure discussed above provides the possibility to introduce bin-wise weights by modifying equation (4.1) to become

$$f_i = \sum_{j=1}^{m} \alpha_j w_{ji} A_{ji}, \tag{4.5}$$

and henceforth using equation 4.5 in the likelihood function $\mathcal{L}$.

The correct shapes of the MC template distributions for the individual sources are obtained by filling histograms with weighted events. In this case the bin contents $a_{ji} = \sum_k w_{kji}$ no longer follow Poisson statistics and the bin errors are given by $\sigma_{ji} = \sqrt{(\sum_k w_{kji}^2)}$ which is in general not equal to the square root of the bin content (the variance of the Poisson distribution). Since this is assumed in the fitting procedure, the use of those templates leads to a wrong result.

In order to properly treat the statistical uncertainties in the fit algorithm, the formalism of equivalent event numbers [21] is adopted and a set of unweighted events that have the same statistical significance as the weighted events is defined. For this purpose, each template histogram is transformed into a template histogram containing equivalent event numbers, $\tilde{a}_{ji}$, and a matching weight value, $\tilde{w}_{ji}$, such that the new histogram multiplied by the weights reproduces the original template content, i.e.

$$\tilde{a}_{ji} \cdot \tilde{w}_{ji} = a_{ji}. \tag{4.6}$$

Furthermore, the equivalent event numbers are required to be Poisson distributed and to have the same relative error as the original event numbers

$$\frac{\tilde{\sigma}_{ji}}{\tilde{a}_{ji}} = \frac{1}{\sqrt{\tilde{a}_{ji}}} = \frac{\sigma_{ji}}{\tilde{a}_{ji}}. \tag{4.7}$$

4.6 Signal Extraction

Equations (4.6) and (4.7) are fulfilled if

$$\tilde{a}_{ji} = \frac{a_{ji}^2}{\sigma_{ji}^2} \quad \text{and} \quad \tilde{w}_{ji} = \frac{\sigma_{ji}^2}{a_{ji}}. \qquad (4.8)$$

The resulting pair of template and weight therefore describes the same distribution and uncertainties as the original template, but obeys Poisson statistics as required for the fit.

2. Full Treatment

In the full treatment the likelihood function is expanded to include weight factors w_{kji} representing the weights of the events of subsample k from source j in bin i. The negative logarithm is then

$$-\ln \mathcal{L} = \sum_{i=1}^{n} d_i \ln f_i - f_i + \sum_{i=1}^{n}\sum_{j=1}^{m}\sum_{k=1}^{l} a_{jki} \ln A_{jki} - A_{jki}, \qquad (4.9)$$

where the number of expected MC events is given by

$$f_i = \sum_{j=1}^{m}\sum_{k=1}^{l} \alpha_j w_{jki} A_{jki}. \qquad (4.10)$$

The solution of the optimization problem is detailed in [15]. It is shown that for any given set of α_j, f_i and A_{jki} can be obtained numerically and that there is a unique solution which makes all A_{jki} positive. The likelihood evaluated with these f_i and A_{jki} then is the function to be minimized by MINUIT. A formal derivation of the equivalent event number method as an approximation to the full treatment can also be found in [15].

4.6.2 Performance of the Fit

The performance of the fit is studied on the basis of the MC samples generated at a center-of-mass energy of $\sqrt{s} = 10\,\text{TeV}$ in 7 $\hat{p}_T$ bins. An example of a fit is shown in Fig. 4.14. The fit is performed using two statistically independent MC subsamples. One sample is used to extract the templates of the simulated shape of the signal and background $p_\perp^{\text{rel}}$ distributions, the other subsample is treated as "data". The plot shows the result obtained from the inclusive sample using the approximate method. The fit result well reproduces the $p_\perp^{\text{rel}}$ shape in "data" and the sample composition agrees within the error. The same statement is true when the full treatment is applied in the fitting procedure.

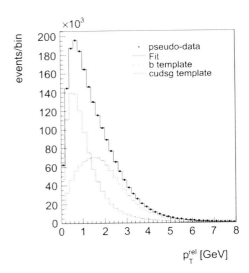

Fig. 4.14 Fit result obtained by dividing the MC sample in two independent subsamples. The *dashed* and the *dotted line* are the *b*- and *cudsg*-template, respectively. The *full circles* correspond to the data distribution, while the *solid line* is the result of the fitting procedure using the approximate method

The dependence on the binning of the input distribution is checked by repeating the fit for different numbers of bins. The fit is found to be stable and the small deviations of the fitted scaling factors are within the statistical errors of the fit.

In order to verify the errors calculated by the fitter and to determine a possible bias in the fit result, repeated fits of pseudo experiments are done. For repeated MC experiments it is not possible to regenerate the full detector simulation. The "truth" is therefore varied from toy experiment to toy experiment to approximately simulate the statistical uncertainty of the templates. This pseudo-truth is obtained by smearing the weighted sum of all templates according to the appropriate error. The data is then generated by Poisson fluctuations of the pseudo-truth. This method takes into account the uncertainty of the MC and the data statistics and thus presents a valid test of the fitting procedure.

In Fig. 4.15 the deviation of the fitted b-fraction from the true value is displayed for repeated fits using the approximate and the full method. In neither of the two methods a significant bias is observed and the width of the pull plots is consistent with one which implies that the errors are calculated correctly by the fitter.

4.7 Validation of MC Templates

4.7.1 Signal

The shape of the signal p_T^{rel} distribution will be validated using a data-driven method. A data sample with a high b-content will be used to cross-check the shape of the

4.7 Validation of MC Templates

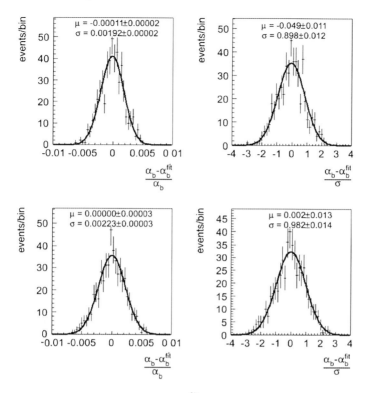

Fig. 4.15 *Left* deviation of the fitted scale factor α_b^{fit} for b-events from the true value α_b in repeated pseudo experiments. *Right* the corresponding pull distributions. The *upper plots* show the result for the approximate fitting method, while the *lower plots* represent the full treatment

$p_\perp^{rel}$ distribution. Such a sample can be obtained for example by selecting events containing single muons with a large impact parameter. This method has been tested in MC. The signed transverse impact parameter of the muon, d_0, is calculated with respect to the jet axis. The impact parameter significance, d_0/σ_{d_0}, is shown in Fig. 4.16 for b-, c- and $udsg$-events. By selecting events with $d_0/\sigma_{d_0} > 30$ a b-purity of 91% is achieved. The efficiency of the selection is 0.8% and a sample with 50,000 events is obtained from data corresponding to an integrated luminosity of 1 pb^{-1}. In MC events, the shape of the $p_\perp^{rel}$ distribution extracted from b-events and from the inclusive sample with a cut on the signed transverse impact parameter significance agree within the statistical uncertainty.

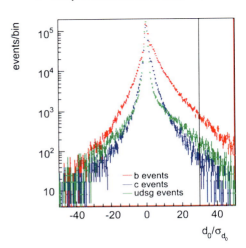

Fig. 4.16 Significance of the signed transverse impact parameter of the muon with respect to the TrackJet direction in signal and background events

4.7.2 Background

The light quark background is dominated by the fake muon contribution and can therefore be estimated from data. If the muon fake probability is known, the light quark background can be measured by re-weighting the hadron spectrum in minimum bias events by the fake muon probability.

While effects of punch-through hadrons mimicking a muon signature mainly become important for high momentum muons, the majority of fake muons with a transverse momentum of $p_T < 30$ GeV is produced in in-flight decays of pions, kaons or baryons. The probability of misidentifying a pion, kaon or proton as a muon is shown in Fig. 4.17 as obtained from simulation. It is a function of the muon transverse momentum and has an average value of about 0.5% for kaons and 0.2% for pions. For protons the muon fake probability is very low and can be neglected in the context of this analysis.

The fake probability of kaons and pions will be measured from data adopting the method described in [22]. Therein the probability that a pion or kaon track is misidentified as a muon is evaluated using $D^0 \to K\pi$ events. A D^0 candidate is reconstructed from any pair of charged tracks with transverse momentum $p_T > 2$ GeV. In order to reduce the combinatorial background, the D^0 meson is required to come from a semileptonic B meson decay, $B^{\pm}/B^0 \to \mu + D^0 + X$. The presence of a muon in the event has the additional benefit of providing a clean signature in the trigger selection. $B \to \mu + D^0 + X$ events are required to fulfill the following event selection:

- global muon with $p_T > 3$ GeV, $-2.1 < \eta < 2.1$, accepted by HLT,
- K candidate: track with $p_T > 2$ GeV and same sign as the muon,
- π candidate: track with $p_T > 2$ GeV and opposite sign to the muon,
- $\Delta R(\mu, K) < 1$ and $\Delta R(\mu, \pi) < 1$,

4.7 Validation of MC Templates

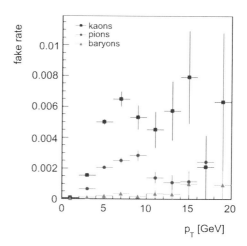

Fig. 4.17 Muon fake probability for kaons, pions and baryons as obtained from the MC simulation

- secondary vertex of K- and π-track with quality $\chi^2 < 10$,
- invariant mass m_{D^0} of the K- and π-candidate in $1.6\,\text{GeV} < m_{D^0} < 2.1\,\text{GeV}$.

The number of signal events is then determined by performing a fit to the invariant mass spectrum of the candidate events. The D^0 mass peak is fitted with a Gaussian function with a mean of 1.865 GeV and a width of 15 MeV. The shape of the background is assumed to be exponential and the parameters are extracted by a fit in the side-bands at $m_{D^0} < 1.8\,\text{GeV}$ and $m_{D^0} > 1.92\,\text{GeV}$. The fake muon probability for pions and kaons is derived using the invariant mass distribution of $D^0 \to K\pi$ events in which one of the decay products is reconstructed as the track belonging to a global muon. In these events the invariant mass spectrum is fitted using the same procedure as described before to determine the number of events in the peak. The fake rate is then given by the number of signal events where one leg is misidentified as a muon divided by the total number of signal events. It can be determined separately for pions and kaons, as the opposite charge allows to distinguish the decay products. The method is tested with the help of a MC sample corresponding to an integrated luminosity of $\mathcal{L} = 0.1\,\text{pb}^{-1}$. In this sample 1,300 D^0 candidates are reconstructed in the signal peak and 5 events with a misidentified kaon are found. These numbers reproduce the kaon fake probability in the MC simulation within the statistical error of 50%. In the case of pions, the available MC statistics is too low to determine the fake probability and it is concluded that a measurement is possible with a data set corresponding to an integrated luminosity of $\mathcal{L} > 0.2\,\text{pb}^{-1}$. The fitted invariant mass distribution for $D^0 \to K\pi$ events is shown in Fig. 4.18 (left), while the subset of events where the kaon is misidentified as a muon is presented in Fig. 4.18 (right).

In order to calculate the muon probability for tracks in an inclusive data sample, the composition of the minimum bias track spectrum has to be known. This information is taken from the MC simulation where the hadronic spectrum consists of 62% pions, 15% kaons and 19% baryons. The fake probability depends on the track transverse

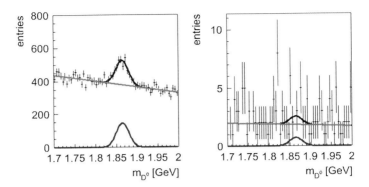

Fig. 4.18 *Left*: fitted invariant mass distribution of D^0 candidates. *Right* fitted invariant mass distribution of D^0 candidates where the kaon leg is misidentified as a muon

momentum. A data set corresponding to a luminosity of $\mathcal{L} > 10\,\text{pb}^{-1}$ would be needed to measure the shape of the transverse momentum dependence from data. As this amount of data will not be available within the time scale of this analysis, we propose to take the shape of the fake probability from simulation and measure the absolute normalization from data.

The scaling of the hadronic track spectrum with the muon fake probability has been performed using MC events. For this study a sample of minimum bias events generated by PYTHIA with a number of events corresponding to an integrated luminosity of $0.2\,\text{nb}^{-1}$ was used. The transverse momentum and pseudorapidity distributions obtained by re-weighting the hadronic track spectrum is compared to the corresponding distributions of fake muons in Fig. 4.19. Furthermore, the $p_\perp^{\text{rel}}$ spectrum calculated from re-weighted tracks and the closest TrackJet is compared to the $p_\perp^{\text{rel}}$ spectrum of fake muons. A good agreement is found for the transverse momentum and the $p_\perp^{\text{rel}}$ spectrum, while the pseudorapidity distributions of fake muons and re-weighted tracks differ significantly. The discrepancy is due to the fact that the fake probability used for re-weighting the hadronic track spectrum is a function of transverse momentum only and the pseudorapidity dependence of the fake probability is not taken into account. This however does not have an effect on the shape of the re-weighted $p_\perp^{\text{rel}}$ spectrum in the simulation. When using the re-weighted $p_\perp^{\text{rel}}$ spectrum as light quark template in the fitting procedure, the resulting b fraction agrees with the fraction obtained from MC within the statistical error. It is concluded that the method of scaling the minimum bias track spectrum is well suited for measuring the light quark background from data.

4.8 Cross Section Measurement   61

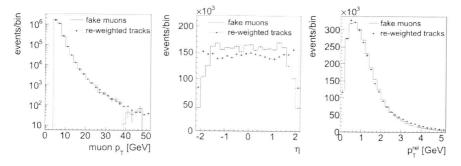

Fig. 4.19 Simulated hadronic track spectrum re-weighted by the fake probability obtained from MC and compared to the simulated fake muon spectrum: transverse momentum (*left*), pseudorapidity (*center*) and $p_\perp^{rel}$ distribution (*right*)

4.8 Cross Section Measurement

4.8.1 Inclusive Cross Section

The inclusive b-quark production cross-section σ is calculated according to the formula

$$\sigma(pp \to b + X \to \mu + X', p_T^\mu > 5 \text{ GeV}, |\eta^\mu| < 2.1) = \frac{\alpha_b \cdot N_{b,rec}^{MC}}{\mathcal{L} \cdot \varepsilon}, \quad (4.11)$$

where α_b is the scale factor for b-events determined by the fit, $N_{b,rec}^{MC}$ is the number of MC events passing the event selection, $\mathcal{L}$ is the integrated luminosity and ε is the efficiency of the trigger and offline selection.

The efficiency of the trigger and offline selection is obtained from the MC simulation as

$$\varepsilon = \frac{N_{b,rec}^{MC}}{N_{b,gen}^{MC}}, \quad (4.12)$$

where $N_{b,gen}^{MC}$ denotes the number of generated b-events in the visible kinematic range. The efficiency is found to be $\varepsilon = 65\%$ where the main factors are the trigger efficiency (79%) and the efficiency of associating a TrackJet to the reconstructed muon (81%).

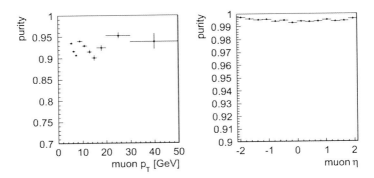

Fig. 4.20 Purity for *b*-events in bins of muon transverse momentum (*left*) and pseudorapidity (*right*). The purity takes into account the contribution from fake muons and resolution effects

4.8.2 Differential Cross Section

The differential *b*-quark production cross-section is measured as a function of muon transverse momentum and pseudorapidity. The flavor composition is determined by performing a fit in each analysis bin. The binning is chosen such that the number of events in every bin is sufficient for a stable fit and at the same time a maximum purity is achieved. The purity $\mathcal{P}$ is defined as the fraction of selected reconstructed events in each bin that have the respective generated quantity in the same bin and is calculated based on the MC simulation:

$$\mathcal{P} = \frac{N_{rec \to gen}^{i,\,MC}}{N_{rec}^{i,\,MC}}. \tag{4.13}$$

Here $N_{rec}^{i,\,MC}$ is the number of reconstructed MC events with muon transverse momentum (or pseudorapidity) in bin i. The reconstrutced muons are matched to generated muons (based on ΔR) in order to determine $N_{rec \to gen}^{i,\,MC}$, i.e. the number of events with both reconstructed and generated muon transverse momentum (pseudorapidity) in bin i. The purity takes into account bin-to-bin migration caused by resolution effects and contributions from fake muons in *b*-events.

The muon transverse momentum range from 5 to 50 GeV is divided into 10 non-equidistant bins with bin sizes between 1 and 20 GeV. The muon pseudorapidity range, from -2.1 to 2.1, is covered by 14 equidistant bins with a bin size of 0.3. A total of 10^3 to 10^5 reconstructed signal events are expected in each bin in a dataset corresponding to 1 pb^{-1} of integrated luminosity. The purity is shown in Fig. 4.20 as a function of muon transverse momentum and pseudorapidity. It is higher than 90% in all bins in muon transverse momentum and above 99% in all bins of muon pseudorapidity.

The templates for the fraction fit are determined individually for each bin. While the $p_\perp^{\rm rel}$ distributions are similar in all bins of muon pseudorapidity, a shift to higher

4.8 Cross Section Measurement

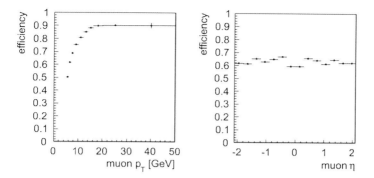

Fig. 4.21 Combined efficiency of the online and offline event selection as a function of muon transverse momentum (*left*) and pseudorapidity (*right*). The efficiency is obtained from simulation

$p_\perp^{\text{rel}}$ values is observed in the bins corresponding to higher muon transverse momenta. The templates and the fit results are shown in Appendix A for all bins.

The differential cross sections $\frac{d\sigma}{dp_T}$ and $\frac{d\sigma}{d\eta}$ are given by

$$\left.\frac{d\sigma}{dp_T}\right|_{\text{bin } i} (pp \to b + X \to \mu + X', |\eta^\mu| < 2.1) = \frac{\alpha_b^i \cdot N_{b,\text{rec}}^{i,\text{MC}}}{\mathcal{L} \cdot \varepsilon^i \cdot \Delta p_T}, \qquad (4.14)$$

$$\left.\frac{d\sigma}{d\eta}\right|_{\text{bin } i} (pp \to b + X \to \mu + X', p_T^\mu > 5 \text{ GeV}) = \frac{\alpha_b^i \cdot N_{b,\text{rec}}^{i,\text{MC}}}{\mathcal{L} \cdot \varepsilon^i \cdot \Delta \eta}, \qquad (4.15)$$

where α_b^i, $N_{b,\text{rec}}^{i,\text{MC}}$ and ε^i refer to the quantities defined in equation (4.11), though evaluated in bin i. Δp_T and $\Delta \eta$ are the widths of the transverse momentum and pseudorapidity bins, respectively. The trigger and offline event selection efficiency in the individual bins is displayed in Fig. 4.21. The efficiency rises as a function of muon transverse momentum from about 50% at $p_T = 5$ GeV to 90% for $p_T > 20$ GeV. The shape of the efficiency as a function of pseudorapidity is explained by the shape of the L1 efficiency of HLT_Mu5.

4.9 Systematic Uncertainties

Due to the high *b*-quark production cross-section, the uncertainty of the measurement will quickly become systematics dominated. In this chapter we cover the main systematic uncertainties that are expected to influence the measured cross section. Some of the systematic uncertainties are expected to be provided by other groups, others will be measured from data directly. The effects of the uncertainty of the detector description as well as the uncertainty on the dynamics of the process (production,

64   4 Study of the Inclusive Beauty Production

fragmentation, decay) are taken into account. In addition, the systematic uncertainty arising from the fitting procedure and the limited number of MC simulations is discussed.

4.9.1 Trigger

In this study we assume an uncertainty of 5% for the L1 and HLT efficiency. An additional systematic uncertainty of 3% is assigned for the muon reconstruction efficiency. The muon trigger efficiency and its uncertainty will be determined from data directly by dedicated study groups using the "tag & probe" method [23]. Therein the known mass resonance of the J/ψ is utilized to select muon candidates and to probe the trigger efficiency. Assuming no efficiency correlation between the two decay products of the resonance it identifies one of the muons (tag) using tight identification criteria and measures the efficiency of the other muon (probe).

4.9.2 Tracking Efficiency and Misalignment

The TrackJet reconstruction algorithm uses all tracks with $p_T > 0.3\,\text{GeV}$ as input. A potential misalignment of the tracking detectors in the early CMS data taking might degrade the performance of the track reconstruction. The actual tracking efficiency might be lower than the tracking efficiency assumed in the detector simulation. The impact of the tracking inefficiency on the TrackJet reconstruction efficiency as well as on the angular resolution is estimated in this study.

The tracking inefficiency is emulated on generator level by discarding a certain fraction of generated charged particles within the tracker volume. The reconstruction efficiency for muons is significantly higher than for pions. Generated muons are thus treated separately and are not rejected in the generated event. The remaining charged particles are used as input to the TrackJet reconstruction algorithm. The reconstruction efficiency and the angular resolution are investigated for different fraction of discarded charged particles corresponding to tracking inefficiencies from 0–20%.

The efficiency for finding a TrackJet close to the reconstructed muon as a function of the tracking inefficiency is shown in Fig. 4.22 (left) for signal events. Since the tracking efficiency has been measured with a precision of 4% [24], a systematic uncertainty of 2% on the number of events which pass the event selection is attributed to the cross-section measurement. The effect of the tracking inefficiency on the jet angular resolution is reflected in the $p_\perp^{\text{rel}}$ distribution (Fig. 4.22, right). In order to get an estimate of the systematic uncertainty of the cross section, the fitting procedure is repeated using the different $p_\perp^{\text{rel}}$ distributions as MC templates for the signal events. The fitted b-fraction deviates by 1% when varying the tracking efficiency within its uncertainty.

4.9 Systematic Uncertainties

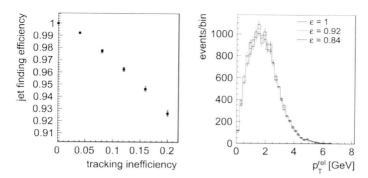

Fig. 4.22 Jet finding efficiency as a function of the tracking inefficiency in signal events (*left*) and shape of the $p_\perp^{rel}$ distribution in signal events for different values of the tracking efficiency (*right*)

4.9.3 Background Composition

The c- and $udsg$-content of the sample cannot be determined separately by the fit. If the c-fraction of the non-b background in data is different from the value used in composing the templates, the fitted b-fraction may vary slightly. The MC simulation predicts a c-fraction of 50–70% in the non-b background depending on the muon transverse momentum. The $udsg$-background is obtained from the inclusive sample and the contribution from fake muons is taken into account. Figure 4.23 shows the results of the fitted b-fraction when varying the c-fraction of the non-b background in the "data" distribution from 0 to 1 while keeping the c-fraction in the background template constant at the value predicted by the MC simulation. The plot is obtained using the statistics of the full samples in the fit. The red lines are at the MC value of the b-fraction in the inclusive sample and the c-fraction in the non-b background.

The uncertainty on the composition can in principle be constrained using data. Non-c-background, in contrast to c-background, will be dominated by fake muons and can therefore be estimated. If the muon fake probability is known, the non-c-background can be measured from data by scaling the hadron spectrum in minimum bias events by the muon fake probability. As discussed in Sect. 4.7, the non-c-background can be obtained from a dataset corresponding to an integrated luminosity of $0.2\,\text{pb}^{-1}$ with an accuracy of 50%. The uncertainty of the c-fraction in the non-b background is indicated by the yellow box in Fig. 4.23. In this region the b-fraction determined by the fitter changes by 3–6% depending on the muon transverse momentum and pseudorapidity.

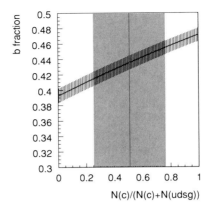

Fig. 4.23 Result of the fitted b-fraction when varying the c-fraction in the background in the "data" distribution while keeping the c-fraction in the background template constant at the value predicted by the MC simulation. The *solid line* is at the MC value of the c-fraction in the background. The *shaded box* indicates the uncertainty of the c-fraction in the background. A systematic uncertainty of the fitted b-fraction of 5% is determined in the inclusive sample

4.9.4 Fragmentation and Decay

About 10% of the b-hadron decays contain a muon from the decay of a c-hadron. These sequential muons carry a smaller fraction of the original b-quark momentum and have a softer momentum and $p_\perp^{rel}$ spectrum. The fraction of sequential muons in the b-sample affects the overall efficiency and the shape of the $p_\perp^{rel}$ spectrum. Varying this fraction within its uncertainty [25] changes the measured cross section by 1%.

Furthermore, the effects of alternative modeling of b-quark fragmentation or b-hadron decays are investigated. The hardness of the fragmentation and the nature of hadronic decays of the b-hadron are expected to have an influence on the transverse momentum spectrum of the muon. Moreover, the description of the fragmentation and the decay might affect the performance of the TrackJet clustering algorithm and consequently the reconstruction efficiency and the angular resolution.

A sample generated with EvtGen is used to investigate the effect of the b-hadron decay properties. The b-quark fragmentation in the PYTHIA sample is modeled by the Peterson fragmentation function and the parameter ϵ_b defines the hardness of the fragmentation. The uncertainty of the fragmentation is studied by varying the parameter ϵ_b between 0.003 and 0.010.

The muon trigger efficiency changes by less than 1% when changing the description of the fragmentation and the decay. This effect can therefore be neglected.

The stability of the TrackJet clustering algorithm is tested by running it on generator level on events produced with the different generator settings. It is found that the efficiency of finding a TrackJet close to the reconstructed muon does not depend on the fragmentation and changes by 1% when simulating the decay with EvtGen.

4.9 Systematic Uncertainties

The relative transverse momentum of the muon with respect to the TrackJet does depend on the modeling of the fragmentation and the decay. In order to quantify the effect, the shapes of the $p_\perp^{\text{rel}}$ distribution obtained from the different samples are again used in the fitting procedure. A change by 1–4% is observed for the value of the fitted b fraction.

The number of $b \to \mu + X$ events inside the acceptance of this analysis depends, in addition to the b-quark production cross-section itself, on the hardness of the fragmentation and the decay branching fractions of b-hadrons. These numbers will be relevant for a comparison with b-quark production calculations but are not systematic uncertainties for a measurement of $\sigma(pp \to b + X \to \mu + X')$.

4.9.5 Production Mechanism

The main production mechanisms that contribute to the $b\bar{b}$ production at LO are flavor creation (hard QCD scattering), flavor excitation (semi-hard process) and gluon splitting (soft process). Since the angle between the two b quarks is smaller in gluon splitting events than in flavor creation or flavor excitation events, the probability of including the two b quarks in the same jet is higher. Thus, the average transverse energy of the b-jet is higher in gluon splitting events. This has an effect on the event selection efficiency as well as on the shape of the $p_\perp^{\text{rel}}$ distribution. The transverse momentum of the muon, the transverse energy of the jet and the $p_\perp^{\text{rel}}$ distribution are shown in Fig. 4.24 for the three production mechanisms.

In the events passing the event selection a fraction of 19% are produced by flavor creation, 56% by flavor excitation and the remaining 25% by gluon splitting. The total event selection efficiency for the three production processes are 62% for flavor creation, 64% for flavor excitation and 67% for gluon splitting events.

The contribution of the three mechanisms to $b\bar{b}$ production is predicted by the MC simulation. To estimate the uncertainty of the relative contributions of the three mechanisms we compare the PYTHIA prediction with the HERWIG prediction and take the difference as the uncertainty. The values obtained in bins of transverse momentum are listed in Table 4.5.

The signal events were then re-weighted in order to adjust the fraction of the three production mechanisms to the value determined by HERWIG. The overall efficiency of the event selection was found to change by less than 1% in all bins. The fitted b-fraction changes by 2–5% depending on the bin in muon transverse momentum.

4.9.6 Description of the Underlying Event

Tracks from the underlying event can change the properties of TrackJets associated to the muon. Especially in events with low multiplicity TrackJets, the distribution of additional tracks may influence the TrackJet reconstruction efficiency and angular

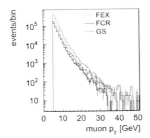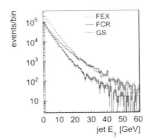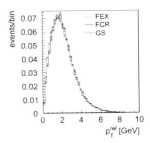

Fig. 4.24 Muon transverse momentum (*left*), TrackJet transverse energy (*center*) and $p_\perp^{rel}$ distribution (*right*) in flavor excitation (FEX), flavor creation (FCR) and gluon splitting (GS) events

Table 4.5 PYTHIA and HERWIG prediction for the relative contributions of the three production mechanisms (flavor creation, flavor excitation, gluon splitting) to the $b\bar{b}$ production

p_T^μ	Flavor creation		Flavor excitation		Gluon splitting	
	PYTHIA	HERWIG	PYTHIA	HERWIG	PYTHIA	HERWIG
5–6 GeV	19%	23%	56%	52%	25%	25%
6–7 GeV	19%	22%	56%	55%	25%	23%
7–8 GeV	19%	20%	56%	55%	25%	25%
8–10 GeV	19%	23%	56%	55%	25%	22%
10–12 GeV	19%	21%	56%	49%	25%	30%
12–14 GeV	18%	26%	55%	48%	27%	26%
14–16 GeV	18%	25%	54%	51%	28%	24%
16–20 GeV	19%	17%	53%	56%	28%	27%
20–30 GeV	18%	20%	52%	39%	30%	41%
30–50 GeV	10%	17%	55%	34%	35%	49%

resolution. The systematic uncertainty due to the description of the underlying event has been studied on the basis of simulated events generated with different MC tunes. They were fit with the standard templates and the observed variation was negligible. The selection efficiency changes are of the order of 10% [26].

4.9.7 Monte Carlo Statistics

The fact that the number of simulated events is limited leads to a systematic uncertainty on the measured *b*-quark production cross-section. The extent of this effect can be estimated by considering the results of the validation of the fitting procedure (Fig. 4.15). An overview of the relative error of the fitted *b*-fraction is presented in Appendix B. The error of the fitted *b*-fraction takes into account the limited MC statistics as well as the limited data statistics.

4.9 Systematic Uncertainties

Table 4.6 Summary of systematic uncertainties

Source	Uncertainty
Trigger	5%
Muon reconstruction efficiency	3%
Tracking efficiency	2%
Background composition	3–6%
Fragmentation	4%
Decay	3%
Production mechanism	2–5%
Underlying event	10%
MC statistics	1–4%
Luminosity	11%
Total	17–18%

4.9.8 Luminosity

The determination of the integrated luminosity and its uncertainty is crucial for this measurement. Different methods are proposed to measure the integrated luminosity at CMS. The quality of the luminosity measurement is a dominate source of systematic uncertainty for the cross section measurement. During the early CMS data taking, the integrated luminosity has been determined with a precision of 11% [27].

4.9.9 Summary

A summary of the main systematic uncertainties of the measurement of the b-quark production cross-section is shown in Table 4.6. The main contribution is due to the uncertainty of the integrated luminosity. The systematic uncertainties depend on the muon transverse momentum and pseudorapidity bin.

4.10 Results

Within the analysis presented here the events of interest are selected by requiring a global muon with transverse momentum $p_T > 5\,\text{GeV}$ and pseudorapidity $|\eta| < 2.1$ and a TrackJet with transverse energy $E_T > 1\,\text{GeV}$ in the reconstructed event.

A sample of 1.8 million selected b-events and about the same number of background events is expected when analyzing proton-proton collision data at a center-of-mass energy of $\sqrt{s} = 7\,\text{TeV}$ corresponding to an integrated luminosity of $1\,\text{pb}^{-1}$. A maximum likelihood fit to the observed $p_\perp^{\text{rel}}$ spectrum based on MC templates of the $p_\perp^{\text{rel}}$ spectrum in signal and background events is performed in order to determine the number of b-events among the selected events.

Table 4.7 Differential b-quark cross-section $d\sigma/dp_T$ for $|\eta| < 2.1$ in bins of muon transverse momentum

p_T^μ (GeV)	N_{sel}	b-fraction	Total efficiency	$d\sigma/dp_T$ [pb]	Systematic %
5–6	1,535,670	0.42±0.002	0.52±0.001	1,229,940	18
6–7	876,872	0.48±0.002	0.62±0.002	671,684	18
7–8	499,809	0.54±0.003	0.69±0.002	391,295	18
8–10	468,886	0.58±0.002	0.76±0.002	178,080	18
10–12	181,433	0.60±0.003	0.81±0.003	67,310	18
12–14	80,323	0.67±0.005	0.86±0.005	31,269	17
14–16	40,137	0.68±0.006	0.89±0.007	15,423	17
16–20	30,486	0.70±0.006	0.92±0.007	5,835	17
20–30	15,413	0.72±0.01	0.90±0.01	1,234	17
30–50	2,796	0.71±0.02	0.94±0.02	106	17

The table shows the number of selected events in 1 pb^{-1}, the b-fraction determined by the fit and total efficiency of the event selection for each bin. In the last two columns the calculated differential cross section as a function of the muon transverse momentum and the systematic uncertainty are given

Table 4.8 Differential b-quark cross-section $d\sigma/d\eta$ for $p_T > 5$ GeV in bins of muon pseudorapidity

η^μ	N_{sel}	b-fraction	Total efficiency	$d\sigma/d\eta$ [pb]	Systematic %
$(-2.1,-1.8)$	209,925	0.52±0.005	0.62±0.004	593,607	17
$(-1.8,-1.5)$	244,033	0.47±0.004	0.61±0.004	629,892	17
$(-1.5,-1.2)$	286,894	0.48±0.004	0.65±0.003	708,336	17
$(-1.2,-0.9)$	277,631	0.49±0.004	0.63±0.003	716,635	17
$(-0.9,-0.6)$	289,568	0.50±0.004	0.65±0.003	744,271	17
$(-0.6,-0.3)$	306,464	0.50±0.004	0.66±0.003	772,191	17
$(-0.3,0)$	261,343	0.51±0.004	0.58±0.003	760,027	17
$(0,0.3)$	261,173	0.53±0.004	0.58±0.003	792,545	17
$(0.3,0.6)$	298,028	0.51±0.004	0.65±0.003	787,942	17
$(0.6,0.9)$	286,553	0.45±0.004	0.63±0.003	679,365	17
$(0.9,1.2)$	279,678	0.47±0.004	0.61±0.003	695,951	17
$(1.2,1.5)$	283,733	0.46±0.004	0.64±0.003	675,545	17
$(1.5,1.8)$	245,298	0.47±0.004	0.62±0.003	617,895	17
$(1.8,2.1)$	210,679	0.47±0.005	0.62±0.004	535,796	17

The table shows the number of selected events in 1 pb^{-1}, the b-fraction determined by the fit and total efficiency of the event selection for each bin. In the last two columns the calculated differential cross section as a function of the muon pseudorapidity and the systematic uncertainty are given

The inclusive b-quark production cross-section in the kinematic range is then calculated by

$$\sigma(pp \rightarrow b + X \rightarrow \mu + X', p_T^\mu > 5 \text{ GeV}, |\eta^\mu| < 2.1) = \frac{\alpha_b \cdot N_{b,rec}^{MC}}{\mathcal{L} \cdot \varepsilon}$$

4.9 Systematic Uncertainties

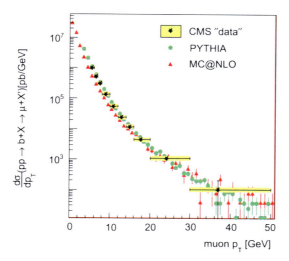

Fig. 4.25 Differential b-quark production cross-section $d\sigma/dp_T$ for $|\eta^\mu| < 2.1$ as a function of the muon transverse momentum. The *black squares* represent the cross section determined by the procedure described in this analysis. The *vertical error bars* show the statistical uncertainty, the systematic uncertainty is indicated by the *yellow area*. The *horizontal error bars* indicated the bin width. The bin center is corrected [28]. The distribution is compared to the prediction of the PYTHIA simulation (*green circles*) and the MC@NLO simulation (*red triangles*)

Fig. 4.26 Differential b-quark production cross-section $d\sigma/d\eta$ for $p_T^\mu > 5\,\text{GeV}$ as a function of the muon pseudorapidity. The *black dots* represent the cross section determined by the procedure described in this analysis. The *error bars* show the statistical uncertainty, the systematic uncertainty is indicated by the *yellow area*. The *horizontal error bars* indicated the bin width. The bin center is corrected [28]. The distribution is compared to the prediction of the PYTHIA simulation (*green line*) and the MC@NLO simulation (*red line*)

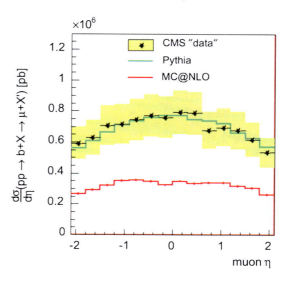

where α_b is the scale factor obtained by the fit that has to be applied to the number of simulated events which pass the event selection $N_{b,rec}^{MC}$, ε is the efficiency of the event selection as determined from simulation and $\mathcal{L}$ is the integrated luminosity. The extrapolation of the measured cross section to the cross section which does not include the definition of the TrackJet amounts to about 10% and is absorbed in the definition of the efficiency. The inclusive cross section can be measured with a negligible statistical uncertainty and a systematic uncertainty of 17%.

In Fig. 4.25 the differential b-quark production cross-section as a function of the muon transverse momentum is shown. The statistical uncertainty is of the order of 1–2%, the systematic uncertainty of the order of 17–18% depending on the muon transverse momentum. In Table 4.7 the factors for calculating the differential b-quark production cross-section as a function of the transverse momentum of the muon are summarized. The results were obtained by dividing the available MC statistics into two independent subsamples using one part to extract the templates and treating the other one as "data".

The result is compared to the leading order and next-to-leading order MC prediction obtained from the PYTHIA and MC@NLO simulation. Since the "data" points correspond to the PYTHIA simulation, they are compatible to the leading order prediction. The MC@NLO distribution lies below the PYHTIA distribution for muon transverse momenta up to about 20 GeV. Above this value the two curves agree within the statistical uncertainty.

The differential b-quark production cross-section as a function of the muon pseudorapidity is shown in Fig. 4.26. The result obtained from "data" matches the PYTHIA prediction. The prediction obtained by the MC@NLO simulation is significantly lower. The systematic uncertainty is of the order of 17% in all bins. The factors for calculating the differential b-quark production cross-section as a function of the muon pseudorapidity are given in Table 4.8.

References

1. DØ Collaboration, Inclusive μ and b-quark production cross sections in $p\overline{p}$ collisions at $\sqrt{s} = 1.8$ TeV. Phys. Rev. Lett. **74**, 3548 (1995)
2. DØ Collaboration, The $b\overline{b}$ production cross section and angular correlations in $p\overline{p}$ collisions at $\sqrt{s} = 1.8$ TeV. Phys. Lett. B **487**, 264 (2000)
3. CDF Collaboration, Measurement of the bottom quark production cross section using semileptonic decay electrons in $p\overline{p}$ collisions at $\sqrt{s} = 1.8$ TeV. Phys. Rev. Lett. **71**, 500–504 (1993)
4. CDF Collaboration, Measurement of the b-hadron production cross section using decays to $\mu^- D^0 X$ final states in $p\overline{p}$ collisions at $\sqrt{s} = 1.96$ TeV. Phys. Rev. D **79**, 092003 (2009)
5. H1 Collaboration, Measurement of beauty production at HERA using events with muons and jets. Eur. Phys. J. C **41**, 453–467 (2005)
6. H1 Collaboration, Measurement of charm and beauty photoproduction at HERA using $D^*\mu$ correlations. Phys. Lett. B **621**, 56–71 (2005)
7. ZEUS Collaboration, Measurement of beauty photoproduction using decays into muons in dijet events at HERA. JHEP **0904**, 133 (2009)
8. ZEUS Collaboration, Measurement of charm and beauty production in deep in elastic ep scattering from decays into muons at HERA. Eur. Phys. J. C **65**, 65–79 (2010)
9. Geant4 Collaboration, Geant4—a simulation toolkit. Nucl. Inst. Meth. A 506 (2003) 250–303
10. CMS Collaboration, CMS High Level Trigger, CERN/LHCC 2007-021 (2007)
11. M. Cacciari, G.P. Salam, G. Soyez, The anti-k_T jet clustering algorithm. JHEP **0804**, 063 (2008)

References 73

12. P. Azzuri et al., Performance of Jet Reconstruction with Charged Tracks Only, CMS AN-2008/041 (2008)
13. P. Schieferdecker et al., Performance of Jet Algorithms in CMS, CMS AN-2008/001 (2008)
14. CMS Collaboration, Plans for Jet Energy Corrections at CMS, CMS PAS JME-07-002 (2007)
15. L. Caminada, W. Erdmann, H.-C. Kästli, Study of the Inclusive b Production Cross-Section in CMS, CMS AN-2009/160 (2009).
16. V. Andreev et al., Measurement of Open Beauty Production at LHC with CMS, CMS AN-2006/120 (2006)
17. R. Barlow, C. Beeston, Comput. Phys. Commun. **77**, 219 (1993)
18. ROOT: an object oriented data analysis framework, http://root.cern.ch
19. http://root.cern.ch/root/htmldoc/TFractionFitter.html
20. http://wwwasdoc.web.cern.ch/wwwasdoc/minuit/minmain.html
21. G. Zech, Comparing Statistical Data to Monte Carlo Simulation—Parameter Fitting and Unfolding, DESY 95-113 (1995)
22. CDF Collaboration, Measurement of correlated $b\bar{b}$ production in $p\bar{p}$ collisions at $\sqrt{s} = 1960$ GeV. Phys. Rev. D **77**, 072004 (2008)
23. D. Bortoletto et al., A Measurement of the Inclusive Production of the Υ Resonance at $\sqrt{s} = 10$ TeV with an Integrated Luminosity of 1 pb^{-1}, CMS AN-2009/118 (2009)
24. CMS Collaboration, Measurement of Tracking Efficiency, CMS PAS TRK-10-002 (2010)
25. C. Amsler et al., Particle Data Group, Phys. Lett. B **667**, 1 (2008)
26. L. Caminada, W. Erdmann, V. Zhukov, M. Niegel, D. Troendle, Measurement of the Cross Section for Open-Beauty Production with Muons and Jets in pp Collisions at $\sqrt{s} = 7$ TeV, CMS AN-2010/171 (2010)
27. CMS Collaboration, Measurement of CMS Luminosity, CMS PAS EWK-10-004 (2010)
28. G.D. Lafferty, T.R. Wyatt, Where to stick your data points: the treatment of measurements within wide bins. Nucl. Instrum. Meth. A **355**, 541–547 (1994)

Chapter 5
Results of First Collisions at $\sqrt{s} = 900\,\text{GeV}$ and $\sqrt{s} = 2.36\,\text{TeV}$

In December 2009 proton-proton collisions at the LHC were recorded with the CMS detector for the first time. The collisions happened at a center-of-mass energy of $\sqrt{s} = 900\,\text{GeV}$ and $\sqrt{s} = 2.36\,\text{TeV}$. The data collected during the first LHC running period were used in this work to study the performance of the physics object reconstruction and to compare it to the results of the MC simulation.

In the first section the event selection is discussed, before an overview of the event simulation is given. The following sections are devoted to the global muon, track and TrackJet reconstruction. A further section addresses the determination of the $p_{\perp}^{\text{rel}}$ variable in data. The conclusions are given in the last section.

5.1 Event Selection

This analysis uses the data collected during runs where the magnetic field was stable at the nominal value of $3.8\,\text{T}$ and the pixel and strip tracking detectors were operational with the high-voltage switched on. Data events corresponding to an integrated luminosity of about $\mathcal{L} = 8.5\,\mu\text{b}^{-1}$ at a center-of-mass energy of $\sqrt{s} = 900\,\text{GeV}$ and about $\mathcal{L} = 0.5\,\mu\text{b}^{-1}$ at $\sqrt{s} = 2.36\,\text{TeV}$ were recorded during these runs.

Because of the relatively low luminosity provided by the LHC during the first running period, the CMS readout was triggered by signals in two elements of the detector monitoring system [1], namely the beam scintillator counter (BSC) [2] and the beam pick-up timing detector (BPTX) [3]. The BSC detectors consist of 16 scintillator tiles and are located at a distance of $\pm 10.86\,\text{m}$ from the nominal interaction point. They cover a pseudorapidity range of $3.23 < |\eta| < 4.65$ and have a time resolution of $3\,\text{ns}$. A more precise timing information with a resolution of $0.2\,\text{ns}$ is provided by the BPTX which are located around the beam pipe at a distance of $\pm 175\,\text{m}$.

The passage of the proton bunches in the beam was detected on the basis of the BPTX signals, while the BSC signals were used to collect minimum-bias collision

L. Caminada, *Study of the Inclusive Beauty Production at CMS and Construction and Commissioning of the CMS Pixel Barrel Detector*, Springer Theses, DOI: 10.1007/978-3-642-24562-6_5, © Springer-Verlag Berlin Heidelberg 2012

76   5   Results of First Collisions at $\sqrt{s} = 900\,\text{GeV}$ and $\sqrt{s} = 2.36\,\text{TeV}$

Table 5.1 Number of data events passing the selection

$\sqrt{s}$	N_{sel}	$\mathcal{L}\,[\mu b^{-1}]$
900 GeV	278179	8.5
2.36 TeV	12932	0.5

events and reject beam background events. Events were accepted if a signal was seen in coincidence in the two BPTXs and if a time coincidence was also recorded in the BSCs compatible with particles coming from the interaction point and incompatible with beam produced particles crossing from one side to the other.

In order to further reduce the background from non-collision events, additional event selection were applied in the offline reconstruction. First, the presence of a reconstructed primary vertex in the event was required. Furthermore, beam-induced background events producing an anomalously large number of pixel hits were excluded by rejecting all events with a fraction of high-purity tracks of less than 0.25 in a track multiplicity larger than 10. The purity of a track is based on the normalized χ^2, the longitudinal and transverse impact parameter and their significance [4]. The number of events passing the event selection are listed in Table 5.1.

5.2 Event Simulation

The MC events were generated with PYTHIA version 6.4 using tune D6T and simulated and reconstructed within the CMS software framework version CMSSW_3_3_6. The event samples consist of inclusive minimum bias events (PYTHIA MSEL $=$ 1 card). The corresponding cross sections and luminosities for proton-proton collisions at $\sqrt{s} = 900\,\text{GeV}$ and $\sqrt{s} = 2.36\,\text{TeV}$ are given in Table 5.2.

The same event selection is applied to MC and data events. The BSC triggers are emulated at the stage of the detector simulation. The number of events which fulfill the event selection criteria are listed in Table 5.2.

5.3 Muon Distributions

A lot of progress has been made in the reconstruction of global muons with the CMS detector using the cosmic muon data collected during two extended data taking periods in autumn 2008 and summer 2009. In the collision data 149 global muons were detected at $\sqrt{s} = 900\,\text{GeV}$ and 15 global muons at $\sqrt{s} = 2.36\,\text{TeV}$. The muons were reconstructed using the standard procedure described in Sect. 2.2.7 and a minimum transverse momentum of $p_T > 1\,\text{GeV}$ and a pseudorapidity of $-2.5 < \eta < 2.5$ were required in the offline selection. Basic distributions of global muons are displayed in Fig. 5.1. There the number of muons per event and the transverse

5.3 Muon Distributions

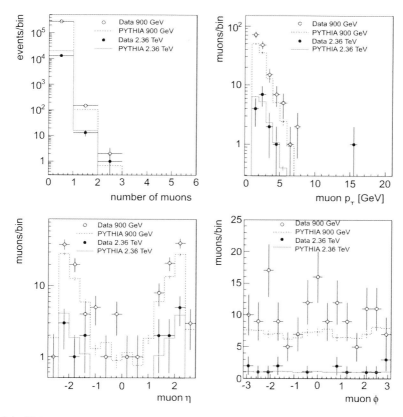

Fig. 5.1 Global muon kinematic distributions in data compared to simulation. The MC distributions have been normalized to the data luminosity. The *open circles* and the *dotted line* correspond to a center-of-mass energy of $\sqrt{s} = 900\,\text{GeV}$ and represent the data and MC events, respectively. The events corresponding to a center-of-mass energy of $\sqrt{s} = 2.36\,\text{TeV}$ are shown as *filled circles* (data) and *solid lines* (MC). *Upper left* Number of muons per event. *Upper right* Muon transverse momentum. *Lower left* Muon pseudorapidity. *Lower right* Muon azimuthal angle

Table 5.2 Overview of the event simulation

$\sqrt{s}$	$\sigma\,[\mu\text{b}]$	N_{gen}	N_{sel}	$\mathcal{L}\,[\mu\text{b}^{-1}]$
900 GeV	52410	10083360	6250310	192.4
2.36 TeV	59960	10654900	6726115	177.7

Events were generated using the PYTHIA MSEL = 1 card. The center-of-mass energy of the proton-proton collisions, the cross section, the number of generated events and the number of selected events are listed. In the last column the equivalent integrated luminosity is given

momentum, pseudorapidity and azimuthal angle spectrum are shown. The shape of the data distributions are in good agreement with the MC prediction.

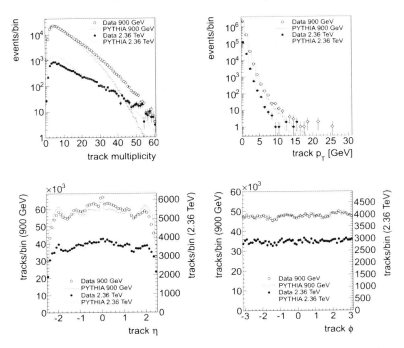

Fig. 5.2 Track kinematic distributions in data compared to simulation. The *open circles* and the *dotted line* correspond to a center-of-mass energy of $\sqrt{s} = 900\,\text{GeV}$ and represent the data and MC events, respectively. The events corresponding to a center-of-mass energy of $\sqrt{s} = 2.36\,\text{TeV}$ are shown as *filled circles* (data) and *solid lines* (MC). *Upper left* Number of tracks per event. *Upper right* Track transverse momentum. *Lower left* Track pseudorapidity. *Lower right* Track azimuthal angle. The simulated distribution of the track multiplicity has been normalized to the data luminosity. Since the track multiplicity is different for data and MC events, the simulated distributions of the kinematic variables are scaled to the number of data events in order to allow for a comparison of the shapes

5.4 Track Distributions

For collision data the track reconstruction was performed using the CTF algorithm (see Sect. 2.2.4) and the tracking detectors had been aligned based on the same procedure as described in [5].

Tracks with a minimum transverse momentum of $p_T > 0.3\,\text{GeV}$ and a pseudorapidity of $-2.5 < \eta < 2.5$ were selected and the quality cuts listed in Table 4.4 were applied. The performance of the track reconstruction for the data is compared to the MC simulation. In Fig. 5.2 the distribution of the track multiplicity and the track transverse momentum, pseudorapidity and azimuthal angle are shown. While the shape of the transverse momentum spectrum in data is well described by MC, the disagreement between data and simulation in track multiplicity and track pseudorapidity is due to an imperfect D6T tuning as discussed in [6]. The asymmetry in the

5.4 Track Distributions 79

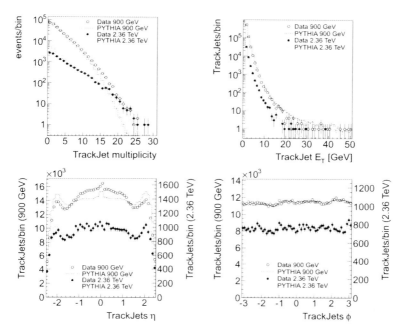

Fig. 5.3 TrackJet kinematic distributions in data compared to simulation. The *open circles* and the *dotted line* correspond to a center-of-mass energy of $\sqrt{s} = 900\,\text{GeV}$ and represent the data and MC events, respectively. The events corresponding to a center-of-mass energy of $\sqrt{s} = 2.36\,\text{TeV}$ are shown as *filled circles* (data) and *solid lines* (MC). *Upper left* Number of TrackJets per event. *Upper right* TrackJet transverse momentum. *Lower left* TrackJet pseudorapidity. *Lower right* TrackJet azimuthal angle. The simulated distribution of the TrackJet multiplicity has been normalized to the data luminosity. Since the TrackJet multiplicity is different for data and MC events, the simulated distributions of the kinematic variables are scaled to the number of data events in order to allow for a comparison of the shapes

azimuthal angle distribution is due to inactive detector modules affecting mainly low momentum tracks. This has been included in the simulation.

5.5 TrackJet Distributions

The tracks selected by the criteria mentioned in the previous section were used as input to the anti-k_T jet clustering algorithm with a cone size of R = 0.5. The reconstructed TrackJets are required to have a transverse energy of $E_T > 1\,\text{GeV}$ and a pseudorapidity of $-2.5 < \eta < 2.5$. An average number of 2.4 TrackJets were reconstructed in the data at $\sqrt{s} = 900\,\text{GeV}$, whereas the average number of TrackJets at $\sqrt{s} = 2.36\,\text{TeV}$ is 3.4. The TrackJet multiplicity is shown in Fig. 5.3. The inadequate description of the track multiplicity by the PYTHIA D6T tune is

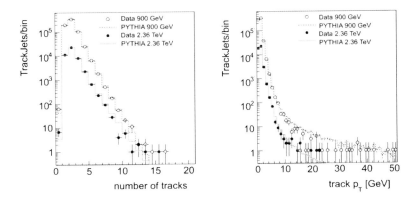

Fig. 5.4 Number of tracks per TrackJet (*left*) and transverse momentum of the highest transverse momentum track in the TrackJets (*right*). The data distributions are compared to simulation. The simulated distributions have been normalized to the number of TrackJets in data. The *open circles* and the *dotted line* correspond to a center-of-mass energy of $\sqrt{s} = 900$ GeV and represent the data and MC events, respectively. The events corresponding to a center-of-mass energy of $\sqrt{s} = 2.36$ TeV are shown as *filled circles* (data) and *solid lines* (MC)

reflected in the TrackJet multiplicity distribution. Figure 5.3 also presents the TrackJet transverse energy, pseudorapidity and azimuthal angle distributions. The shape of the transverse momentum is well described by simulation and an average transverse energy of $E_T = 1.8\,(1.4)$ GeV in data (MC) at $\sqrt{s} = 900$ GeV and $E_T = 1.9\,(1.9)$ GeV at $\sqrt{s} = 2.36$ TeV is measured.

In Fig. 5.4 the number of tracks reconstructed within a TrackJet and the transverse momentum of the highest transverse momentum track are compared to the MC simulation and a good agreement is found. These results are relevant in view of the measurement of the *b*-quark production cross-section as the analysis presented here is based on a precise determination of the muon momentum with respect to the TrackJet direction. A good understanding of the TrackJet reconstruction and a reliable simulation of the TrackJet distributions are thus of utmost importance.

Furthermore, the performance of the reconstruction of the transverse impact parameter was investigated. In the measurement of the *b*-quark cross section events with muons with a large transverse impact parameter significance will be used in order to obtain a *b*-enriched sample. Alternatively, it might be possible to use the transverse impact parameter of the tracks in the TrackJet as discriminating variable. In Fig. 5.5 the transverse impact parameter significance of all tracks in the TrackJet and of the track with the highest transverse impact parameter significance is shown. The impact parameter is calculated with respect to the reconstructed primary vertex and the sign is relative to the axis of the TrackJet. The distributions of the MC simulations are in good agreement with the data distributions.

5.6 $p_\perp^{rel}$ Distribution

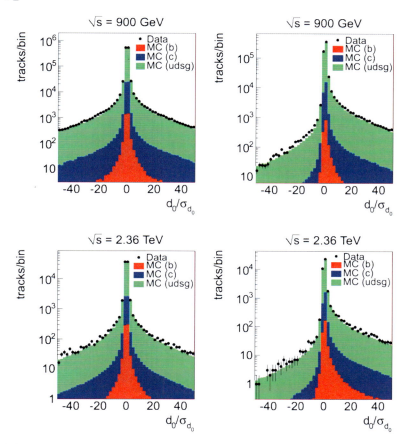

Fig. 5.5 Transverse impact parameter significance of all tracks in the TrackJet (*left*) and of the track with the highest transverse impact parameter in the TrackJet (*right*). The *upper plots* correspond to a center-of-mass energy of $\sqrt{s} = 900$ GeV, the *lower plots* to $\sqrt{s} = 2.36$ TeV. The simulated distributions are normalized to the number of tracks in data. The *black circles* represent the data distribution. The MC distributions are split into the contribution of b (*red*), c (*blue*) and $udsg$ (*green*) events

5.6 $p_\perp^{rel}$ Distribution

According to the PYTHIA simulation, the b-quark production cross section is $\sigma_b = 28\,\mu$b at $\sqrt{s} = 900$ GeV and $\sigma_b = 96\,\mu$b at $\sqrt{s} = 2.36$ TeV. Thus, the number of events containing b-quarks is very low in the 2009 collision data and a measurement of the cross section using the $p_\perp^{rel}$ method will not be possible. Nonetheless it is instructive to study the $p_\perp^{rel}$ distribution in this data in order to better understand the background.

The $p_\perp^{rel}$ distribution in the data at $\sqrt{s} = 900$ GeV is shown in Fig. 5.6. The $p_\perp^{rel}$ variable is determined in events with a muon with transverse momentum

Fig. 5.6 $p_\perp^{rel}$ distribution in the collision data at $\sqrt{s} = 900\,\text{GeV}$ compared to the MC simulation. The *black circles* represent the data distribution. The MC distributions are divided into the contribution of b (*red*), c (*blue*) and $udsg$ (*green*) events

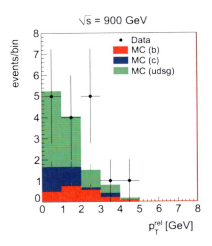

$p_T > 3\,\text{GeV}$ and pseudorapidity $-2.1 < \eta < 2.1$ and a TrackJet with transverse momentum $E_T > 1\,\text{GeV}$. In data, 16 events pass this selection, whereas in simulation only 11 events are selected. In the data at $\sqrt{s} = 2.36\,\text{TeV}$ two events are selected and values of $p_\perp^{rel} = 1.45\,\text{GeV}$ and $p_\perp^{rel} = 1.49\,\text{GeV}$ are measured. These numbers are in agreement with the simulation where also 2 events are reconstructed and a mean value of $p_\perp^{rel} = 1.36\,\text{GeV}$ is found.

The method for validating the background templates by re-weighting the hadronic track spectrum by the muon fake probability (see Sect. 4.7) is investigated. The data statistics is too low to perform a measurement of the muon fake probability and it is therefore taken from simulation. For muons with transverse momentum $p_T < 5\,\text{GeV}$ the fake probability strongly depends on the pseudorapidity as can be seen in Fig. 5.7.

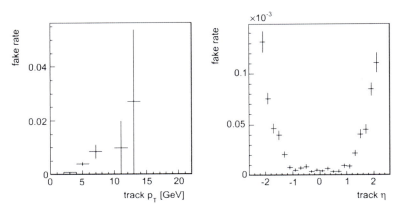

Fig. 5.7 Muon fake probability as a function of transverse momentum (*left*) and pseudorapidity (*right*) as obtained from simulation

5.7 Conclusions

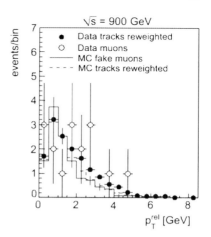

Fig. 5.8 $p_\perp^{\text{rel}}$ distribution obtained by re-weighting the hadronic track spectrum in data by the simulated muon fake probability (*filled circles*). The result is compared to the simulated fake muon spectrum (*solid line*), the re-weighted hadronic track spectrum in MC (*dotted line*) and the measured $p_\perp^{\text{rel}}$ distribution in data (*open circles*). The MC distributions are scaled to the data luminosity

When re-weighting the track spectrum in minimum bias events the dependence of the fake probability on the transverse momentum and the pseudorapidity is taken into account. The result is shown in Fig. 5.8.

A harder $p_\perp^{\text{rel}}$ distribution is observed in the re-weighted data with respect to simulation. In order to validate the weighting procedure, it was applied to the simulated track spectrum. The results from simulation agree within the statistical error. For comparison, the measured $p_\perp^{\text{rel}}$ distribution calculated from muons which were most likely generated in light quark or charm decays is also shown in Fig. 5.8.

5.7 Conclusions

The data recorded during the first collisions in December 2009 have been used to study the performance of the physics object reconstruction. The number of global muons in the collision data is very small. However, the reconstruction of global muons was already well commissioned using cosmic data. The reconstruction of tracks and TrackJet was studied and a in general the data were found to be well described by simulation. This result is most valuable in view of the analysis presented in this thesis.

The $p_\perp^{\text{rel}}$ distribution obtained in the data corresponding to $\sqrt{s} = 900$ GeV was determined and compared to the simulation. The MC simulation is in agreement with data although the statistics is very limited. Furthermore, the method for obtaining the light-quark background $p_\perp^{\text{rel}}$ templates using a data-driven approach was validated.

References

1. CMS Collaboration, The CMS experiment at the CERN LHC, JINST **3**, S08004 (2008)
2. A.J. Bell, Design and construction of the beam scintillation counter for CMS, Master's Thesis, University of Canterbury, New Zealand, 2008
3. T. Aumeyr, Beam phase and intensity monitoring for the compact Muon solenoid experiment, Master's Thesis, Vienna University of Technology, Austria, 2008
4. CMS Collaboration, Tracking and Vertexing Results from First Collisons, CMS PAS TRK-10-001 (2010)
5. CMS Collaboration, Alignment of the CMS Silicon Tracker during Commissioning with Cosmic Rays, JINST **5**, T03009 (2010)
6. CMS Collaboration, Measurement of the Underlying Event Activity in Proton-Proton Collisions at 900 GeV, CMS PAS QCD-10-001 (2010)

Chapter 6
Preliminary Results of First Collisions at $\sqrt{s} = 7\,\text{TeV}$

On March 30, 2010, the first proton-proton collisions at a center-of-mass energy of $\sqrt{s} = 7\,\text{TeV}$ happened at the LHC. The data statistics recorded by the CMS detector during the first months of data-taking allows for a first measurement of the inclusive b-quark production cross-section at the LHC [1]. The preliminary result has been presented at the 35th International Conference on High Energy Physics [2].

In this chapter the analysis of the collision data at $\sqrt{s} = 7\,\text{TeV}$ collected in April and May, 2010 is summarized. In the first section an overview of the event simulation is given, followed by a short discussion of the event selection and the signal extraction. The preliminary results together with the main systematic uncertainties are summarized and discussed in the last two sections.

6.1 Event Simulation

The CMS data are compared to the PYTHIA MC simulation version 6.4 with tune D6T. For simulation and reconstruction the CMS software version CMSSW_3_5_X was used. Two statistically independent samples were generated: an inclusive QCD minimum bias sample and a muon-enriched QCD sample, in which the presence of a generated muon with $p_T > 2.5\,\text{GeV}$ and $|\eta| < 2.5$ was required (compare to Sect. 4.2). An overview of the MC simulation is given in Table 6.1.

6.2 Event Selection

The event selection detailed in Sect. 4.5 is applied to the CMS data and the MC simulation.

L. Caminada, *Study of the Inclusive Beauty Production at CMS and Construction and Commissioning of the CMS Pixel Barrel Detector*, Springer Theses, DOI: 10.1007/978-3-642-24562-6_6, © Springer-Verlag Berlin Heidelberg 2012

Table 6.1 Overview of the PYTHIA event simulation at $\sqrt{s} = 7\,\text{TeV}$. An inclusive QCD minimum bias sample and a muon-enriched QCD sample were generated

Data set	$\sqrt{s}$	$\sigma\,[\mu b]$	ϵ_{filt}	N_{reco}	$\mathcal{L}\,[\text{nb}^{-1}]$
inclusive QCD	7 TeV	71260	1	10998457	0.15
muon enriched QCD	7 TeV	48440	0.0018	10418911	119.5

The center-of-mass energy of the proton-proton collisions, the cross section, the filter efficiency of the generator level filter and the number of selected events are listed in the table. In the last column the equivalent integrated luminosity is given.

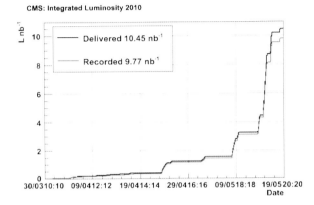

Fig. 6.1 Integrated luminosity delivered by the LHC and recorded by CMS during the first data taking at $\sqrt{s} = 7\,\text{TeV}$ in April and May, 2010

6.2.1 Run Selection

The data recorded during the first months of CMS data taking at $\sqrt{s} = 7\,\text{TeV}$ between March 30 and May 19, 2010 were used for this analysis. The integrated luminosity as a function of time is shown in Fig. 6.1.

In the analysis only runs certified by the Data Quality Monitoring group were considered. The run selection is based on the following criteria:

- Stable beam conditions with beam energy at $E = 3.5\,\text{TeV}$
- Stable magnetic field inside CMS with $B_z > 3.7\,\text{T}$
- Goodness of L1 trigger and HLT
- Silicon pixel and strip tracking detectors in readout with nominal high-voltage settings
- DT, CSC and RPC muon detectors in readout with nominal high-voltage settings

The data recorded during the runs fulfilling the quality criteria correspond to an integrated luminosity of $\mathcal{L} = 8.1\,\text{nb}^{-1}$.

6.2 Event Selection 87

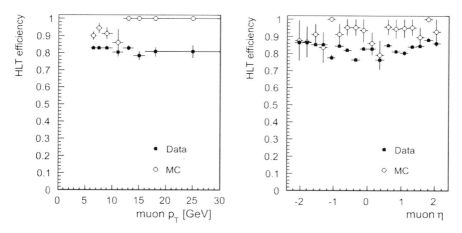

Fig. 6.2 Single muon trigger efficiency (HLT_Mu3) as measured from data compared to simulation. The simulated trigger efficiency is determined on the basis of inclusive QCD MC events. *Left*: HLT efficiency as a function of the muon transverse momentum for muons with pseudorapidity $|\eta| < 2.1$. *Right*: HLT efficiency as a function of muon pseudorapidity for muons with transverse momentum $p_T > 6\,\text{GeV}$

6.2.2 Trigger Selection

The events of interest are selected by the HLT_Mu3 single muon trigger path (see Sect. 4.3). The trigger efficiency is measured from data in minimum bias events. The minimum bias events were obtained as described in Sect. 5.1 and the HLT_Mu3 efficiency is calculated by

$$\varepsilon_{HLT} = \frac{N^\mu_{reco,HLT}}{N^\mu_{reco}}, \tag{6.1}$$

where N^μ_{reco} is the number of reconstructed global muons with transverse momentum $p_T > 6\,\text{GeV}$ and pseudorapidity $|\eta| < 2.1$. $N^\mu_{reco,HLT}$ is the number of reconstructed global muons in the same kinematic range and accepted by the HLT.

The trigger efficiency as a function of muon transverse momentum and pseudorapidity is shown in Fig. 6.2. It should be noted that the MC simulation overestimates the trigger efficiency.

6.2.3 Offline Selection

Background from non-collision events is reduced by requiring a reconstructed primary vertex with more than three tracks while beam-induced background events were rejected if the fraction of high-purity tracks to all tracks was less than 0.25.

Table 6.2 Selection criteria for the reconstructed global muons in the offline analysis

Variable	Value
minimum transverse momentum	$p_T > 6\,\text{GeV}$
maximum pseudorapidity	$|\eta| < 2.1$
longitudinal impact parameter	$z_0 < 20\,\text{cm}$
transverse impact parameter	$d_0 < 5\,\text{cm}$
number of pixel layers with hits	$\geqslant 2$
number of valid hits in tracker	$\geqslant 12$
number of valid hits in muon chambers	> 0
normalized χ^2 of tracker track	< 10
normalized χ^2 of global track	< 10

Events with a reconstructed global muon with transverse momentum $p_T > 6\,\text{GeV}$ and pseudorapidity $|\eta| < 2.1$ are selected. In order to improve the background rejection, quality criteria on the number of hits in the tracker and muon detector, the χ^2 of the track fit and the impact parameter are imposed. The selection criteria are listed in Table 6.2. In Fig. 6.3 the transverse momentum, pseudorapidity, azimuthal angle and transverse impact parameter distributions for selected muons in data and simulation are shown. The data distributions are compared to the muon-enriched QCD MC sample. The MC distribution are normalized to the data luminosity and the simulated trigger efficiency is corrected with respect to the trigger efficiency measured in data. The shape of the muon transverse momentum and azimuthal angle distribution are well described by MC whereas in the pseudorapidity distribution discrepancies between data and MC are observed.

The b-jet is defined as the TrackJet containing the muon. After subtracting the muon momentum from the TrackJet momentum, the TrackJet energy is required to fulfill $E_T > 1\,\text{GeV}$. The probability of associating a TrackJet to the reconstructed muon is 77.4 ± 0.3 and $80.8 \pm 0.1\%$ in data and simulation, respectively. A total of 16826 data events pass the selection.

The distribution of the TrackJet kinematic variables are compared to simulation and the result is displayed in Fig. 6.4. In general, the data distribution are well described by simulation. The TrackJet pseudorapidity distribution is more central since it is correlated with the pseudorapidity distribution of the muon which also shows this feature. The disagreement between data and MC in the TrackJet transverse energy distribution at low transverse energy is due to the imperfect PYTHIA D6T tune which does not describe correctly the track multiplicity and the track transverse momentum spectrum for low transverse momentum tracks [3].

6.3 Signal Extraction

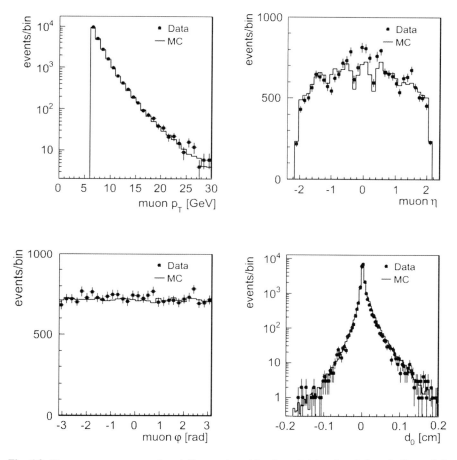

Fig. 6.3 Transverse momentum (*top left*), pseudorapidity (*top right*), azimuthal angle (*lower left*) and transverse impact parameter (*lower right*) distribution of selected reconstructed global muons. The data distribution (*circles*) is compared to the MC simulation (*solid line*) normalized to the integrated luminosity

6.3 Signal Extraction

6.3.1 Data-Driven Determination of Light Quark Background

The light quark background template is obtained from data using the method introduced in Sect. 4.7. Hadrons satisfying all muon track selection criteria (except for muon identification) are re-weighed with the muon fake probability and used instead of muons to determine the $p_\perp^{\text{rel}}$ template. The muon fake probability is taken from MC

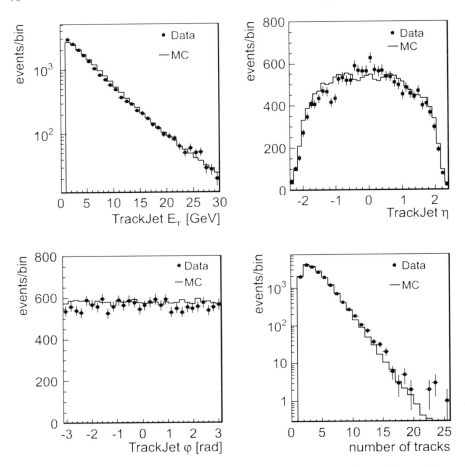

Fig. 6.4 Transverse energy (*top left*), pseudorapidity (*top right*), azimuthal angle (*lower left*) and number of tracks (*lower right*) of selected TrackJets. The data distribution (*circles*) is compared to the MC simulation (*solid line*) normalized to the integrated luminosity

simulation, as the current data sample size does not allow a precise determination of this quantity.

The $p_\perp^{\text{rel}}$ distribution of re-weighted tracks in minimum bias events is compared to the one of simulated muons in light quark events in Fig. 6.5. The $udsg$ template determined from data is harder than in simulation. In order to evaluate the systematic uncertainty of the cross-section measurement due to an imperfect background description, the $p_\perp^{\text{rel}}$ distribution obtained from data as well as from simulation is used in the fitting procedure.

6.3 Signal Extraction

Fig. 6.5 $p_\perp^{rel}$ distribution in light quark background events obtained by re-weighting the hadronic track spectrum in data by the simulated muon fake probability (*circles*). The result is compared to the simulated fake muon spectrum (*solid line*) and the re-weighted hadronic track spectrum in simulation (*dotted line*). The MC distributions are normalized to the number of data events

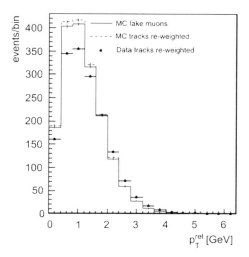

6.3.2 Data-driven Validation of $p_\perp^{rel}$ Templates in Signal Events

The shape of the signal $p_\perp^{rel}$ distribution is validated using a data-driven method (see Sect. 4.7). A data sample with a high b-content is used to cross-check the shape of the $p_\perp^{rel}$ distribution. Such a sample is obtained by selecting events containing single muons with a large impact parameter. By selecting events with $d_0/\sigma_{d_0} > 20$ a b-purity about 90% is achieved. The shape of the $p_\perp^{rel}$ distribution extracted from data events and MC events (b and inclusive) with a cut on the muon signed transverse impact parameter significance are shown for the inclusive sample in Fig. 6.6.

No systematic uncertainty is introduced as the $p_\perp^{rel}$ distribution in signal events obtained from simulation agrees with the data distribution within the limited statistics.

6.3.3 $p_\perp^{rel}$ Fit

The fitting procedure, discussed in detail in Sect. 4.6, is carried out to determine the fraction of b-events among the selected signal events. A binned maximum likelihood to the observed $p_\perp^{rel}$ distribution based on templates obtained from simulation (signal and part of the background) and data (the remaining background) is performed. The fitting procedure is applied to the $p_\perp^{rel}$ spectrum in the inclusive sample, 8 bins in muon transverse momentum and 7 bins in muon pseudorapidity. The result of the fit in the inclusive sample is displayed in Fig. 6.7, while the results in bins of muon transverse momentum and pseudorapidity are shown in Appendix C.

Fig. 6.6 $p_\perp^{rel}$ distribution in b-enriched events in data and MC obtained by a cut on the muon signed transverse impact parameter significance $d_0/\sigma_{d_0} > 20$. The *circles* are the data events, the *solid line* corresponds to the b-events in MC and the *dotted line* is the inclusive QCD MC. The distribution are normalized to the number of data events

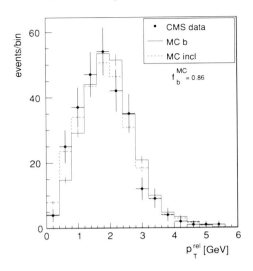

Fig. 6.7 $p_\perp^{rel}$ distribution measured in data. Overlaid are the result of the maximum likelihood fit and the simulated template distributions. The *dashed* and the *dotted line* are the b- and $cudsg$-template, respectively. The filled circles correspond to the data distribution, while the *solid line* is the result of the fitting procedure

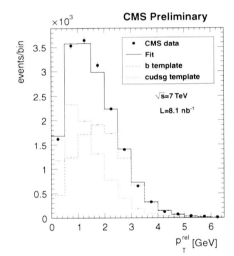

In Table 6.3 the b-fractions determined by the fit in the different bins and their statistical uncertainty are listed. Furthermore, the likelihood ratio χ^2 of the fits is given where the number of degrees of freedom of all fits is 10.

The stability of the fit is tested by performing repeated fits with varied binning. The results agree within the statistical error.

6.4 Results

Table 6.3 b-fraction determined by the fit for the inclusive sample and the bins in muon transverse momentum and pseudorapidity

	Fitted b-fraction	χ^2
Inclusive		
	0.44 ± 0.01	28.1
p_T^μ		
6–7 GeV	0.42 ± 0.02	9.1
7–8 GeV	0.39 ± 0.03	5.2
8–10 GeV	0.48 ± 0.03	23.0
10–12 GeV	0.57 ± 0.04	12.9
12–14 GeV	0.45 ± 0.06	8.8
14–16 GeV	0.45 ± 0.09	6.6
16–20 GeV	0.54 ± 0.10	6.5
20–30 GeV	0.78 ± 0.15	5.6
η^μ		
$(-2.1, -1.5)$	0.42 ± 0.04	11.2
$(-1.5, -0.9)$	0.40 ± 0.03	6.1
$(-0.9, -0.3)$	0.46 ± 0.03	5.4
$(-0.3, 0.3)$	0.44 ± 0.03	17.4
$(0.3, 0.9)$	0.47 ± 0.03	18.1
$(0.9, 1.5)$	0.49 ± 0.03	6.8
$(1.5, 2.1)$	0.43 ± 0.04	17.4

The errors represent the statistical uncertainty. In addition, the likelihood-ratio χ^2 of the fit is given. The number of degrees of freedom in the fit is 10.

6.4 Results

The inclusive b-quark production cross-section is calculated according to equation (4.11):

$$\sigma^{vis} \equiv \sigma(pp \to b + X \to \mu + X', p_T^\mu > 6\,\text{GeV}, |\eta^\mu| < 2.1) = \frac{\alpha_b \cdot N_{b,rec}^{MC}}{\mathcal{L} \cdot \varepsilon}.$$

The efficiency ε includes the trigger efficiency (82%), the muon reconstruction efficiency (97%), and the efficiency for associating a TrackJet to the reconstructed muon (77%). The trigger efficiency is determined from data, the other two efficiencies are taken from MC simulation.

The preliminary result of the inclusive b-quark production cross-section within the kinematical range is

$$\sigma(pp \to b + X \to \mu + X', p_T^\mu > 6\,\text{GeV}, |\eta^\mu| < 2.1)$$
$$= (1.48 \pm 0.04_{\text{stat}} \pm 0.22_{\text{syst}} \pm 0.16_{\text{lumi}})\,\mu\text{b}.$$

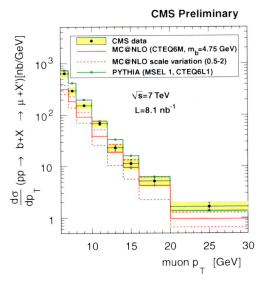

Fig. 6.8 Differential b-quark production cross-section $d\sigma/dp_T$ for $|\eta^\mu| < 2.1$ as a function of the muon transverse momentum. The *black* symbols show the measured cross section where the vertical error bars represent the statistical uncertainty and the horizontal error bars indicate the bin width. The *yellow* band shows the quadratic sum of statistical and systematic errors. The systematic error (11%) of the luminosity measurement is not included. The *dashed red lines* illustrate the MC@NLO theoretical uncertainty as described in the text. The solid green line shows the PYTHIA result

where the error represents the statistical error of the fit. The systematic error is discussed in the following section.

For comparison, the inclusive b-quark production cross section predicted by the PYTHIA and MC@NLO is given:

$$\sigma^{vis}_{\text{PYTHIA}} = 1.8\,\mu\text{b},$$
$$\sigma^{vis}_{\text{MC@NLO}} = [0.84^{+0.36}_{-0.19}(\text{scale}) \pm 0.08(m_b) \pm 0.04(\text{pdf})]\,\mu\text{b}.$$

The error for MC@NLO is obtained by changing the QCD renormalization and factorization scales independently from half to twice their default values within a fiducial volume as in [4]. The massive HERWIG calculation agrees with the MC@NLO prediction within the theoretical uncertainties.

The preliminary results of the differential b-quark production cross-section as a function of muon transverse momentum and pseudorapidity are shown in Figs. 6.8 and 6.9 and summarized in Tables 6.4 and 6.5.

6.5 Systematic Uncertainties

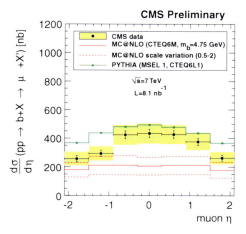

Fig. 6.9 Differential b-quark production cross-section $d\sigma/d\eta$ for $p_T^\mu > 6\,\text{GeV}$ as a function of the muon pseudorapidity. The *black* symbols show the measured cross section where the vertical error bars represent the statistical uncertainty and the horizontal error bars indicate the bin width. The *yellow* band shows the quadratic sum of statistical and systematic errors. The systematic error (11%) of the luminosity measurement is not included. The *dashed red* lines illustrate the MC@NLO theoretical uncertainty as described in the text. The *solid green* line shows the PYTHIA result

Table 6.4 Differential b-quark cross-section $d\sigma/dp_T$ for $|\eta^\mu| < 2.1$ in bins of muon transverse momentum

p_T^μ	N^b	ε	$d\sigma/dp_T$ [nb/GeV]	stat	sys	lumi
6–7 GeV	2897 ± 140	0.56 ± 0.01	640	5%	15%	11%
7–8 GeV	1479 ± 96	0.61 ± 0.01	297	7%	15%	11%
8–10 GeV	1674 ± 93	0.67 ± 0.01	154	6%	14%	11%
10–12 GeV	771 ± 58	0.69 ± 0.02	68	7%	12%	11%
12–14 GeV	282 ± 38	0.76 ± 0.02	23	14%	13%	11%
14–16 GeV	135 ± 27	0.73 ± 0.04	11	20%	14%	11%
16–20 GeV	131 ± 25	0.78 ± 0.04	5.2	19%	12%	11%
20–30 GeV	102 ± 20	0.77 ± 0.04	1.6	19%	11%	11%

The number of b-events (N^b) determined by the fit, the efficiency (ε) of the online and offline event selection, and the differential cross section together with its relative statistical, systematic and luminosity uncertainty are given.

6.5 Systematic Uncertainties

The systematic uncertainties of this analysis have been discussed in detail in Sect. 4.9 and are summarized in the following. The dominant contributions are due to the description of the $udsg$ background and of the underlying event. The modeling of b-quark production, semileptonic b-hadron decays, and the signal efficiency is better understood and has less impact on the systematic error.

Table 6.5 Differential b-quark cross-section $d\sigma/d\eta$ for $p_T^\mu > 6\,\text{GeV}$ in bins of muon pseudorapidity

η^μ	N^b	ε	$d\sigma/d\eta$ [nb]	stat	sys	lumi
$(-2.1,-1.5)$	773 ± 68	0.62 ± 0.02	256	9%	16%	11%
$(-1.5,-0.9)$	895 ± 71	0.63 ± 0.02	293	8%	15%	11%
$(-0.9,-0.3)$	1322 ± 84	0.64 ± 0.02	424	6%	15%	11%
$(-0.3,0.3)$	1240 ± 82	0.59 ± 0.02	434	7%	14%	11%
$(0.3,0.9)$	1333 ± 84	0.64 ± 0.02	426	6%	14%	11%
$(0.9,1.5)$	1119 ± 75	0.61 ± 0.02	375	7%	14%	11%
$(1.5,2.1)$	802 ± 66	0.63 ± 0.02	262	8%	14%	11%

The number of b-events (N^b) determined by the fit, the efficiency (ε) of the online and offline event selection, and the differential cross section together with its relative statistical, systematic and luminosity uncertainty are given.

The muon trigger efficiency has been determined from data in minimum bias events. The statistical uncertainty on the trigger efficiency amounts to 3–5%, depending on the muon transverse momentum and pseudorapidity, and is taken as a systematic uncertainty. The muon reconstruction efficiency is known to a precision of 3%. The tracking efficiency for hadrons is known with a precision of 4%. This induces a systematic uncertainty of 2% on the number of events passing the event selection. The uncertainty in the tracking efficiency affects the b-fraction in the fit by about 1%.

The background template consists of contributions from $c\bar{c}$ events and from light quark events, where a hadron is misidentified as a muon. The fit does not separately determine the c and $udsg$-content of the sample. Two effects can introduce a systematic error. (i) The $udsg$ template determined from data could be biased. Using the PYTHIA-derived $udsg$ template introduces a difference to the nominal fit of 1–10%, depending on the muon transverse momentum and pseudorapidity bin. (ii) If the c-fraction of the non-b background in the data were different from the value used in composing the templates, the fitted b-fraction would change somewhat. The error due to the background composition amounts to 3–6%.

The uncertainty of the contributions of the $b\bar{b}$ production mechanisms, the modeling of the b-quark fragmentation and the b-hadron decay lead to systematic uncertainties of 2–5, 4 and 3%, respectively. The systematic error due to the modeling of the underlying event is of the order of 10%.

At the present early stage of the CMS experiment, the integrated luminosity recorded is known to about 11%.

6.6 Conclusions

A preliminary result of the inclusive b-quark production cross-section in proton-proton collisions at a center-of-mass energy of $\sqrt{s} = 7\,\text{TeV}$ has been determined for the first time within this thesis. The result is based on data corresponding to an

6.6 Conclusions

integrated luminosity of $\mathcal{L} = 8.1\,\mathrm{nb}^{-1}$ recorded by the CMS detector during the first months of data taking in April and May, 2010.

The determination of the cross section relies on the performance of the muon and track reconstruction. The muon and TrackJet kinematical distributions were compared to the MC prediction and the simulation was found to be in good agreement with data. Furthermore, A detailed study of the dominant sources of systematic uncertainty has been performed in the scope of this work.

The measured b-quark production cross section was compared to LO and NLO MC predictions. The data tends to be higher than the MC@NLO prediction at low transverse momentum and central rapidity.

References

1. CMS Collaboration, Open beauty production cross section with muons in pp collisions at $\sqrt{s} = 7$ TeV, CMS PAS BPH-10-007 (2010)
2. L. Caminada, Measurement of the inclusive b production cross section in pp collisions at $\sqrt{s} = 7$ TeV, 35th International Conference on High Energy Physics (ICHEP 2010)
3. CMS Collaboration, Tracking and Primary Vertex Results in First 7 TeV Collisions, CMS PAS TRK-10-005 (2010)
4. M. Cacciari, S. Frixione, M.L. Mangano et al., QCD analysis of first b cross-section data at 1.96 TeV. JHEP **07**, 033 (2004)

Part II
Construction and Commissioning of the CMS Pixel Barrel Detector

Chapter 7
The CMS Pixel Barrel Detector

The CMS pixel detector allows for high precision tracking in the region closest to the interaction point in a particularly harsh environment characterized by a high track multiplicity and heavy irradiation. The main purpose of the pixel detector is the reconstruction of secondary vertices from heavy flavor and tau decays and the generation of track seeds for track reconstruction.

The barrel part of the CMS pixel detector was developed, designed and built at the Paul Scherrer Institute in cooperation with ETH Zürich and the University of Zürich. In this chapter, the main components of the CMS pixel barrel (BPIX) detector are introduced. An overview of the detector design and the mechanical structure is given, followed by a detailed description of the detector module and its main building blocks. In the last section, the readout and control system of the BPIX detector is explained.

7.1 Design of the CMS Pixel Barrel Detector

The CMS BPIX Detector [1] is composed of three cylindrical layers at mean radii of 4.4, 7.3 and 10.2 cm and has a length of 53 cm. It is built from 768 highly segmented silicon sensor modules with a pixel size of $100 \times 150 \, \mu m^2$ providing about 48 million readout channels (Fig. 7.1). The pixels are almost square shaped as emphasis has been put on achieving a similar track resolution in both the $r\phi$ and z direction. The pixel detector layers are divided into two half shells built from 0.25 mm thin carbon fiber ladders. Each ladder is glued to an aluminum cooling pipe with 0.3 mm wall thickness and holds 8 sensor modules. In order to make the layers hermetic in the $r\phi$ plane, the ladders are mounted with overlap as can be seen in Fig. 7.1. The edge ladders in the region where two half shells meet are designed as half ladders equipped with half modules to reach full spatial coverage. Three half shells are mounted together at the endflange and form one half of the BPIX detector.

The BPIX detector is attached to four 2.2 m long supply tubes which carry the services along the beam pipe and house the electronics for the detector readout and

L. Caminada, *Study of the Inclusive Beauty Production at CMS and Construction and Commissioning of the CMS Pixel Barrel Detector*, Springer Theses, DOI: 10.1007/978-3-642-24562-6_7, © Springer-Verlag Berlin Heidelberg 2012

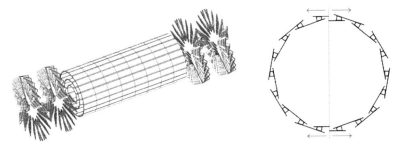

Fig. 7.1 *Left* Illustration of the mechanical design of the three barrel layers and the four endcap disks of the CMS pixel detector. *Right* radial cut through the mechanical frame of the first barrel layer

control. In addition, the supply tube accommodates the cooling lines which feed the 10 cooling loops of the barrel detector mechanics. The supply tubes are a complex system in design as well as in production due to the thin radial shell thickness (1–2 cm), the large number of circuits, plugs and sensors, and the fine wires and thin printed circuit boards that were used. The detector and the supply tubes are connected via a six layer PCB which is mounted on the detector endflange and distributes the power and the control signals to the individual modules. The final BPIX system consists of two independent half cylinders, an inner (+x) and an outer (−x), with a detector half shell and two supply tubes.

7.2 Detector Modules

The sensitive element of a module is a silicon sensor [2] with a dimension of 66.6×18.6 mm^2 and a thickness of about $285\,\mu$m. An array of 16 or 8 readout chips (ROCs) [3] for full and half modules, respectively, is bump-bonded to the sensor. The ROC dimension is 8×8 mm^2 and each ROC is segmented into 4,160 pixel readout channels. On the other side of the silicon sensor a three layer high density interconnect (HDI) flex printed circuit is glued and wire bonded to the ROCs. A token bit manager chip (TBM) [4], that controls the readout of the ROCs, is mounted on top of the HDI. To fix the module to the mechanical support structure two base strips made of $250\,\mu$m thick silicon nitride (Si$_3$N$_4$) are glued to the ROC side of the module. A power cable consisting of 6 copper coated aluminum wires is soldered to the HDI and brings analog, digital and high voltage to the module. The control and readout signals are sent through a two layer Kapton signal cable which is wire-bonded to the HDI. The HDI distributes the signals and the voltages to the ROCs.

The size of a full module is 66.6×26 mm^2 and the weight is up to 3.5 g depending on the length of the signal and power cables. The average power consumption of a full module is 2 W. Figure 7.2 shows all components of a BPIX detector module.

7.2 Detector Modules

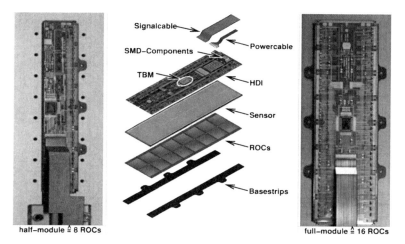

Fig. 7.2 Picture of a BPIX half module (*left*) and full module (*right*). *Center* the components of a pixel barrel detector module (from *top* to *bottom*) the Kapton signal cable, the power cable, the HDI, the silicon sensor, the 16 ROCs and the base strips

7.2.1 Sensor

When a charged particle passes through the sensor, electron-hole pairs are produced in the sensor material which are separated by an electric field. In the CMS pixel detector the electrons are collected at the anode and passed through an indium bump to the ROC. A minimum ionizing particle crossing the sensor at an angle of 90° creates an average ionization charge of about 22,000 electrons. The CMS pixel detector silicon sensor adopts a double sided processed n+ on n design. High dose n+ pixels are implanted in a high resistance n substrate (Fig. 7.3). A moderated p-spray technique is used for interpixel isolation. The backside is p-doped forming the pn junction. After irradiation the substrate material will be type inverted and the depletion will start at the structured n-side. Therefore a sensor design was chosen that allows partially depleted operation at very high fluence. A non-irradiated sensor can be operated at a bias voltage of 100–150 V while for irradiated sensors voltages up to 600 V have to be applied to achieve full depletion. A stable operation at these very high bias voltages is possible due to a multiple guard ring structure on the p-side that allows to keep all sensor edges on ground potential.

7.2.2 Readout Chip

Each silicon sensor pixel segment is electrically connected via an indium bump to a ROC pixel unit cell (PUC). The PUCs are arranged in 26 × 80 double columns which are controlled by the double column periphery. The double columns, the double

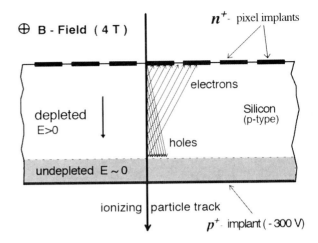

Fig. 7.3 Diagram of a charged particle crossing the sensor of the BPIX module [5]. The charge introduced by the passage of the ionizing particle is collected at the high dose n+ implants

column periphery and the chip periphery are the three main functional units of a ROC. They fulfill the task of recording the position and charge of all hit pixels with a time resolution of 25 ns and store the information on-chip during the L1 trigger latency. The behavior of the ROC is controlled by means of 26 digital-to-analog converter (DAC) registers which can be programmed using a modified I^{2}C interface running at 40 MHz.

The PUC can receive a signal either through a charge deposition in the sensor or by injecting a calibration signal. Within the PUC, the signal is first passed through a two stage pre-amplifier/shaper system to a comparator where zero-suppression is applied. The comparator threshold is set by a DAC for the whole ROC but can be adjusted via a 4-bit DAC (trim bits) for each pixel individually. If a signal exceeds the comparator threshold the hit information is stored, the corresponding pixel becomes insensitive and the column periphery is notified. The column periphery writes the value of the bunch crossing counter into a time stamp buffer and issues a readout token. A column drain mechanism is initialized to read out the pixel hit information. Hit pixels send the registered analog pulse-height information together with the pixel address to the column periphery before being set again into data taking mode.

The double column periphery verifies the trigger by comparing the time stamp with a counter running behind the bunch crossing counter by the trigger delay. In case of agreement the column is set into readout mode and the data acquisition is stopped, otherwise the data are discarded. When the readout token arrives at the double column periphery the validated data are sent to the chip periphery and the double column is reset. The ROCs are read out serially via a 40 MHz analog link. A picture of a BPIX readout chip is shown in Fig. 7.4.

7.2 Detector Modules

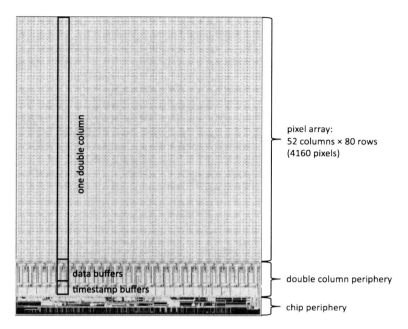

Fig. 7.4 Picture of the BPIX readout chip highlighting the three main building blocks: double column, double column periphery and chip periphery [6]

7.2.3 Token Bit Manager

A TBM chip is wire-bonded to the HDI and controls the readout of the ROCs. Since there are two analog data links per BPIX module to the FED for the inner two layers, the TBM is configured as pairs in a dual TBM chip. The main functionality of the TBM is to synchronize the data transmission. For each incoming L1 trigger, the TBM issues a readout token. The token is passed to the ROCs and the readout is initialized. The last ROC in the chain sends the token back to the TBM. The TBM multiplexes the signal from the ROCs, adds a header and a trailer to the data stream and drives the signal through the readout link. In addition, the TBM distributes the L1 trigger and the clock to the ROCs. A schematic of the readout chain is displayed in Fig. 7.5.

7.3 Readout and Control System

The BPIX readout system is organized into 64 independent readout groups consisting of analog and digital opto-hybrid circuits which serve 8, 12 or 16 modules and provide communication between the detector and the front-end modules in the underground

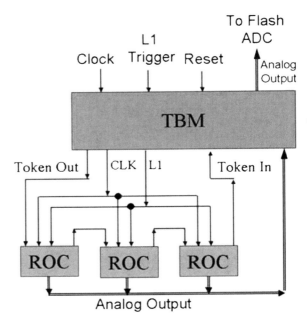

Fig. 7.5 Schematics of a readout chain consisting of a TBM and a group of ROCs [4]

service room. The electronics for the readout and control system are integrated on the detector supply tube. A supply tube is divided into 8 sectors which contain the power lines and the readout and control electronics of two readout groups, one serving the modules of the first two layers, the other serving the modules of the third layer.

Figure 7.6 presents an overview of the pixel readout and control system. A detailed description can be found in [7]. The system provides a very complex functionality and consists of three main parts: an analog read-out link from the BPIX modules to the front-end drivers (FED), a digital control link from the pixel front-end controller (pxFEC) to the modules and a slow control link from a standard front-end controller (FEC) to the supply tube to configure the readout electronics hosted on the supply tube. The individual components of the pixel readout and control system are described in more detail in the following.

7.3.1 Analog Chain

An example of an analog readout signal of a module with a single pixel hit is shown in Fig. 7.7. The data stream contains the readout signals from each ROC in the readout group preceded by the TBM header and ended by the TBM trailer. The TBM header uses eight clock cycles and starts with three very low analog levels called ultra-black (UB) and a black level which defines the zero level of the differential analog signal. The additional four levels contain the event number information. The

7.3 Readout and Control System

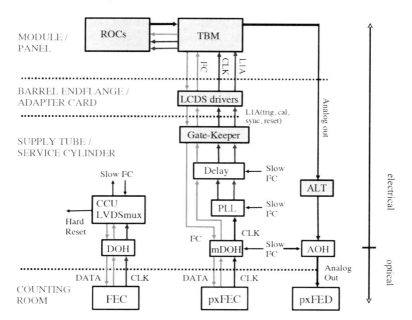

Fig. 7.6 Overview of the BPIX readout and control system

readout sequence of a ROC begins with a UB level, a black level and a level called last DAC which comprises the value of the DAC addressed by the last programming command. The hit information is transmitted in 6 clock cycles encoding the pixel address and the pulse height information. The pixel address consists of the pixel row and column number coded in six discrete analog levels. The TBM trailer consists of two UB levels and two black levels followed by four clock cycles transmitting the TBM error status.

The data stream which contains all hit information belonging to a single trigger is sent out by the TBM through the module Kapton cable. The Kapton cable consists of differential analog lines separated by quiet lines from the lines for the fast digital signals. The analog signals are split from the digital signals on the endring PCB. A single Kapton cable brings the analog signals of one readout group to the printed circuit board on which the Analog Optical Hybrids (AOHs) [8] are placed. The electric analog signals are amplified in an Analog Level Translator (ALT) chip and converted into 40 MHz analog optical signals in the AOHs. Each AOH is equipped with 6 lasers which drive the signal through optical fibers to the front end drivers.

7.3.2 Front End Driver

A total of 32 front end drivers (FEDs) [9] are setup in two 9U VME crates located in the CMS underground service room. The optical fibers connecting the AOHs to

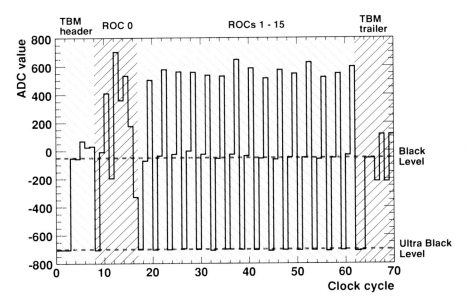

Fig. 7.7 Analog readout of a full module with one hit in ROC 0

the FEDs have to cover a distance of about 60 m. A FED has 36 optical inputs each equipped with an optical receiver and an ADC. The FED receives the analog data, digitizes the signals at the LHC frequency and decodes the pixel address information. It then builds event fragments and sends them to the central DAQ system. Alternatively the FED can be operated in a transparent mode making unprocessed ADC output data available for calibration and testing purpose. A programmable offset voltage can be set for each optical input in order to compensate for bias shifts in the analog signal.

7.3.3 Supply Tube

Four supply tube half cylinders hold the readout and control circuits of the pixel detector. Each sector of the supply tube includes an analog opto-board with 6 AOHs and a digital opto-board with two digital opto-hybrids (DOHs) [10] on the detector near side. The front-end controller modules send the clock and trigger information together with other control signals through the DOHs to the detector. A DOH is connected to four optical fibers, two for receiving and two for sending signals. The LHC clock and trigger information is encoded in one signal which is sent over a single fiber to the DOH. A phase locked loop (PLL) chip [11] is used to split the clock from the trigger information before sending it to the detector modules. The relative phases of all control signals are adjusted using a DELAY25 chip [12]. A Gatekeeper chip converts the low voltage differential signal (LVDS) used by the

7.3 Readout and Control System

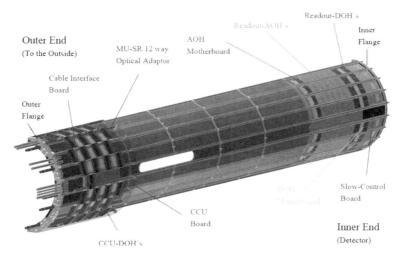

Fig. 7.8 Schematic view of one of the four BPIX supply tubes

PLL and the DELAY25 chip to a low current differential signal (LCDS) used by the pixel front-end chips. The electronic devices needed in the digital control circuit are mounted on the digital opto-boards. In each sector 44 optical fibers are needed for communication with the front-end modules, 36 for the analog readout and 8 for the digital control.

A communication and control unit (CCU) board is placed in the central sector of the supply tube. It is the core component developed for slow control, monitoring and timing distribution and is described in more detail in the next section. The slow control signals like temperatures, pressures and humidities are also brought together in the central slot and connected by dedicated slow control adapter boards to the readout cables. A schematic view of a supply tube half cylinder is shown in Fig. 7.8.

7.3.4 Communication and Control Unit

The pixel detector front end control system consists of four CCU boards equipped with 9 CCU chips [13]. Each board supervises one quarter of the detector. The slow control links are implemented as a ring architecture. A ring consists of 9 CCUs, two optical drivers and receivers that bring clock, trigger and control data to the CCUs and a front-end controller (FEC) [14] which is the master of the network. The CCUs distribute the digital control signals to the individual readout boards in each sector. A CCU chip supports two I^{2}C channels to communicate with the front-end readout electronics, and three PIA channels to generate the necessary signals to reset the

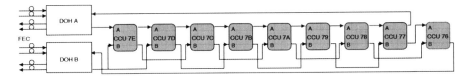

Fig. 7.9 Layout of the BPIX CCU board showing the CCU chips, the DOHs and the doubled interconnection lines. The doubled signal paths (called A and B) and the ability of bypassing of interconnection lines ensures the high operational stability of the system

AOHs, DOHs and ROCs of one sector. Eight CCUs are used for the control of the eight sectors, the ninth CCU is a dummy CCU used for redundancy.

Since a sizable number of front-end channels depend on the same control link, a very high reliability of the system is of utmost importance. A CCU failure leads to a loss of communication to all electronics attached to the CCU. A redundancy scheme based on doubling signal paths and bypassing of interconnection lines, between the CCUs and between the CCUs and the FEC, is supported. The dummy CCU allows to mitigate a single DOH failure. The CCU is equipped with two DOHs which form separated control rings and thus ensure a high operational reliability. The DOHs on the CCU board are programmed by the first two CCU chips. The layout of the CCU board is shown in Fig. 7.9.

7.3.5 Front End Controller

The pixel detector uses two different types of front-end controllers: a Pixel FEC (pxFEC) and a standard Tracker FEC (trFEC). In both cases, 8 mezzanine daughter cards are mounted on the FEC mother cards. The pxFEC is used to program the detector front-end chips (TBMs, ROCs) and to distribute the clock and trigger signal to the modules. The trFEC provides communication with the CCUs and the programming of the front-end devices (AOHs, DOHs, PLL, DELAY25). While the trFEC uses the standard I^2C protocol, a pixel specific modified I^2C protocol is implemented in the pxFEC.

The FEC performs a transmission verification by comparing the number of bytes sent to the number of bytes returned by the front-end and checking the returned hub/port address. Status bits with the result of the comparison are set by the FEC and stored for possible review. The transmission verification by the FEC is not used in standard operation mode.

References

1. H.C. Kästli et al., CMS barrel pixel detector overview. Nucl. Instrum. Meth. A **582**, 724 (2007)
2. Y. Allkofer et al., Design and performance of the silicon sensors for the CMS barrel pixel detector. Nucl. Instrum. Meth. A **584**, 25 (2008)

References

3. H.C. Kästli et al., Design and performance of the CMS pixel detector readout chip. Nucl. Instrum. Meth. A **565**, 188 (2006)
4. E. Bartz, The token bit manager chip for the CMS pixel readout, in *Proceeding of the Ninth Workshop on Electronics for LHC Experiments*, CERN 2003-006, CERN/LHCC/2003-055 (2003)
5. CMS Collaboration, The Compact Muon Solenoid: Technical Proposal, CERN-LHCC-94-38 (1994)
6. C.M.S. Collaboration, The CMS experiment at the CERN LHC. JINST **3**, S08004 (2008)
7. D. Kotlinski et al., The control and readout systems of the CMS pixel barrel detector. Nucl. Instrum. Meth. A **565**, 73 (2006)
8. Analog Optohybrids, Technical Specifications, http://aoh.hephy.at
9. M. Pernicka et al., The CMS Pixel FED, in *Proceedings of the Topical Workshop on Electronics for Particle Physics*, pp. 487–491, Prague, Czech Republic, 2007
10. J. Troska et al., IEEE Trans. Nucl. Sci. **NS-40**(4), 1067 (2003)
11. P. Placidi et al., CMS Tracker PLL Reference Manual, CERN Document Server
12. H. Correia et al., Delay25—A 4 Channel 1/2 ns Programmable Delay Line, CERN-EP/MIC, No. 14 (2000)
13. C. Paillard et al., in *Proceedings of the Eighth Workshop on Electronics for LHC Experiments*, CERN 2002-003, CERN/LHCC/2002-034 (2002)
14. K.A. Gill et al., in *Proceedings of the Eleventh Workshop on Electronics for LHC Experiments* (*LECC 2005*), CERN-2005-011, CERN-LHCC-2005-038 (2005)

Chapter 8
Construction and Commissioning of the CMS Pixel Barrel Detector

The CMS BPIX detector system was assembled and fully tested at PSI before it was transported to CERN [1]. A slice of the CMS control and data acquisition system has been setup at PSI in order to have a tool for operating and testing larger segments of the pixel detector.

A main focus of the hardware related work during this thesis was on the setup of the test system at PSI and the development of software algorithms needed for testing, calibrating and monitoring all detector components during the integration at PSI and later during the installation phase at CERN. In addition, this work contributed to the successful completion of the construction, commissioning and installation of the final detector system.

In the first two sections the development of the testing procedure is described, before the construction of the BPIX detector is addressed. In particular the technique for mounting the modules on the support structure, the assembly of the detector control and readout electronics on the supply tube and the integration of the final system are explained. The following section focusses on the installation of the BPIX detector into the CMS detector. In the last section the results of the system tests are discussed and the performance of the BPIX during the first running period is reviewed. The chapter concludes with a summary.

8.1 Low Level Hardware Testing Procedure

The goal of this work was to develop a reliable and fast testing procedure in view of the commissioning of the BPIX detector at PSI and the installation of the detector inside CMS. Due to the tight schedule of the final integration of CMS, a well established procedure which allows to determine in a very short period of time whether the central elements of the detector were working was of utmost importance during the installation. Furthermore emphasis was placed on the realization of a standalone testing procedure which would not depend on the availability of the standard CMS

L. Caminada, *Study of the Inclusive Beauty Production at CMS and Construction and Commissioning of the CMS Pixel Barrel Detector*, Springer Theses, DOI: 10.1007/978-3-642-24562-6_8, © Springer-Verlag Berlin Heidelberg 2012

114   8 Construction and Commissioning of the CMS Pixel Barrel Detector

software framework. The implementation of the standalone software allows direct communication with the main hardware components and is thus very well suited to quickly identify potential failures.

8.1.1 Test Setup

The BPIX control and data acquisition system set up at PSI was built from front-end modules equivalent to the ones which will be used during data-taking at CERN. Initially, a group of 12 detector modules were operated (System12). The modules were connected through an endring print to an AOH board equipped with 2 AOHs which transmit the analog optical signals to the FED. A pxFEC, a trFEC and a fully equipped DOH board formed the digital detector control. The System12 setup was mainly used to test the individual components of the readout system and to gain operational experience with the very complex system. Furthermore, it presented a useful tool for testing and debugging new software developments.

A commissioning system with the true mechanical structure and prototype supply tubes was assembled before the construction of the final detector. In the commissioning system only two sectors were equipped with modules. A second test stand was established which allowed to operate the commissioning system independent of System12. This test stand was later used to commission the two halves of the final detector system. The final detector system was tested sector-wise since only one sector could be connected to the readout at a time due to the limited number of optical fibers and FED modules available.

8.1.2 Software Architecture

The BPIX readout and control system consists of three main parts (see Sect. 7.3): the analog readout link between the modules and the FED, the digital control link between the modules and the pxFEC and the slow control link between the CCUs and the trFEC. At P5 the FED and FEC boards are installed in different VME crates in the CMS underground service room. Since one crate of electronic boards is controlled by one PC, the software used for communication is implemented as a distributed system.

The standalone BPIX online software has been developed using the socket technology [2] provided by the framework of Python [3]. The supervisor processes— called trFECProcess, pxFECProcess and FEDProcess—are implemented in C++ and communicate with the trFEC, pxFEC and FED VME boards, respectively. Similarly, the PSProcess controls the CAEN power supply. The clock and trigger information is sent from the TTC board which is programmed using a standard CMS XDAQ [4] process.

8.1 Low Level Hardware Testing Procedure

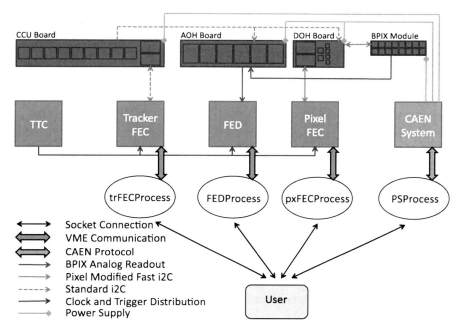

Fig. 8.1 Schematic view of the BPIX low level hardware test software architecture

Through the Python socket interface the user can send messages to the control processes and receive the answers (one query at a time). In Fig. 8.1, a schematic overview of the different software components is shown. In a standard testing sequence, first the clock and trigger distribution is started and the CAEN power supply is turned on. Then the delay settings are adjusted by issuing commands in the trFECProcess, the BPIX modules TBMs and ROCs are programmed and the readout is initialized. In the end, the analog data is read out via the FEDProcess and the digital readback is verified using the pxFECProcess.

For the offline data analysis the ROOT [5] framework was used. In addition, a ROOT program called FEDScope operated as a digital oscilloscope to monitor the FED output.

8.1.3 Testing Sequence

The test procedure is implemented as a bottom-up approach and is divided into four stages which are discussed in the following.

116   8 Construction and Commissioning of the CMS Pixel Barrel Detector

1. Functionality of the detector control and communication system

First, the proper working of the CCU ring architecture, the redundancy mechanism and the correct programming of the I^2C devices is verified. In addition, the hard resets issued by the trFEC and sent through the CCU reset line to the AOHs, the DOHs and the ROCs are tested. The AOH and DOH resets can be checked by reading the values in the chip registers. Since at this point of the testing procedure communication with the detector modules is not yet established, an alternative method has to be used to spot the effect of the ROC reset. This can be done by monitoring the power consumption via the PSProcess as the digital current is increased by a hard reset.

2. Performance of the analog readout chain

If the tests of the detector and the control and communication system prove successful, the test sequence is continued, from now on investigating one readout group at a time. In a first step the mapping between the AOH and the FED channels is checked. Each AOH is equipped with 6 lasers for which the bias and the gain can be adjusted individually. The optical fibers are combined in groups of 12 which then connect to one FED input. A FED has three inputs and serves the two readout groups of one sector. The mapping is verified by increasing the bias of a given laser and checking the light intensity received in the corresponding FED channel. In case a wrong match is found the problem is resolved by either redoing the connection on the supply tube or updating the original map. In a next step, the noise and the gain on the analog transmission line is measured by performing a scan of the full laser bias range. As an example the light intensity, the slope and the noise as a function of the laser bias are shown for one AOH channel in Fig. 8.2. Poor or dirty connections strongly affect the light transmission since light might be either absorbed (which lowers the gain) or scattered back to the laser (which increases the noise). Channels with a noise value of more than 4 ADC counts or very low gain could in most cases be improved by re-cleaning the optical connections.

3. Performance of the digital readback mechanism

In order to establish I^2C communication with the BPIX modules the delays in the digital readout circuit have to be adjusted. The phase of each digital transmission line can be independently programmed in steps of 0.5 ns from 0 to 25 ns using the corresponding register in the Delay25 chip. Additionally, the PLL chip implements a function that generates 12 different clock phases evenly distributed between 0 and 12 ns.

In CMS the LHC clock and the L1 trigger decision are transmitted from the counting room to the detector using one single fiber. To achieve this, both the clock and trigger signals are encoded as a single signal as schematically illustrated in Fig. 8.3. At the receiving end the signal is decoded by the PLL chip and sent via two separate lines, LHC clock (CLK) and Calibrate/Trigger/Reset (CTR), through the Delay25 chip to the BPIX modules. In addition, the CLK signal is split in the

8.1 Low Level Hardware Testing Procedure

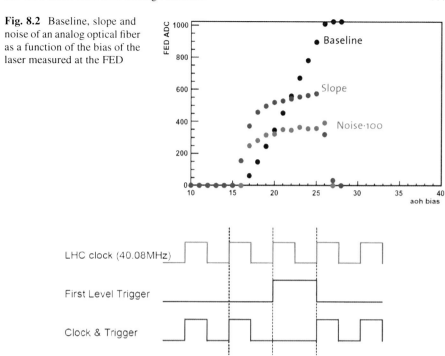

Fig. 8.2 Baseline, slope and noise of an analog optical fiber as a function of the bias of the laser measured at the FED

Fig. 8.3 Combined coding of the LHC clock and L1 trigger signals [6]. The encoding is done such that the coded signal is identical to the clock signal if the L1 trigger issues a reject decision. If however an event is accepted by the L1 trigger, the coded signal stays at the low level for the duration of one clock cycle

Gatekeeper chip and one line (RCK) is returned and sent again through the Delay25 chip. The digital programming and control data (SDA) also goes through the Delay25 chip. If the gate is open the SDA is transmitted to the BPIX modules which sends the acknowledge signal (RDA) back, otherwise the data packet is returned in the gatekeeper. The digital circuit is illustrated in Fig. 8.4.

Since the Delay25 and the PLL chip serve all modules of the corresponding readout group, they cannot be used to compensate for delays between modules caused by different lengths of their signal cables. For this purpose further delays can be added on the endring PCB.

The SDA signal can only be decoded by the TBM if it is in phase with the CLK signal, so that the start and stop conditions are recognized correctly. The working region for sending data is common to all modules in a readout group and is determined by performing a scan of the Delay25 SDA register whilst the delay of the CLK signal is kept at a constant value. In each point of the scan, the value of the DAC

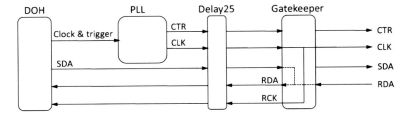

Fig. 8.4 Illustration of the BPIX digital circuit

register that regulates the analog voltage used by the ROC is increased for all ROCs and the analog current is watched via the PSProcess. If the current increases, the programming is successful and a valid delay setting is found. It should be mentioned that this method is time consuming as the current reading on the power supply is stable only after about 3 s. The plateau of the correct phase is about 6–8 ns and a working point in the middle of the plateau is chosen.

4. TBM and ROC programming and module analog readout

After an operation point for module programming is found, the digital readback of the modules is tested. If a programming command is received by a module, the TBM sends an acknowledge sequence which includes the TBM address. The TBM addresses take values between 0 and 31 and are assigned according to a predefined address scheme within a readout group. The address assignment is checked by going through a scan of the RCK delay: programming commands are sent to the modules and the returned addresses are decoded in the pxFEC. An example of a RCK scan is shown in Fig. 8.5.

The RCK delay range in which the acknowledge of a module is received has a length of about 10 ns and depends on the module position on the detector mechanics since the phase shift between RCK and RDA signal is due to different cable lengths and the delay added on the endring print. The RCK delay however is set globally for the whole readout group. In order to determine a working point common to most modules in a readout group, a two dimensional scan of RCK delay versus RDA delay is performed. The setting of the RCK and RDA delay does not influence the data taking as the digital readback is only a diagnostic tool. It is chosen such that the readback of a maximum number of modules is received and it is re-adjusted if needed.

Almost all readout groups showed satisfactory results when testing the readback. The most common failures were caused by bad connections either of the module cables at the PCB or between the PCB and the supply tube and could be recovered.

Besides the information about the relative displacement of the RCK delay setting range between the modules in a readout group was used to measure the signal speed in the Kapton signal cable. Since the length of the signal cables and their position on

8.1 Low Level Hardware Testing Procedure

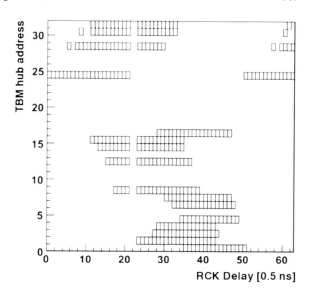

Fig. 8.5 Digital readback of 16 modules in a readout group as a function of the RCK delay. The boxes indicated the RCK values at which the programming commands were received and returned correctly by the TBM. The plateau of correct phase is about 10 ns. It is shifted among the modules in a readout group since the signal path depends on the position of the module on the detector mechanics

the endring print were known, the signal speed could be calculated. The measured value of 7 ± 2 ns/m is in good agreement with the technical specification of 5.39 ns/m.

In the last stage of the low level hardware tests the functionality of the modules is verified. The trigger signal is sent to the modules to initialize the readout and the analog signal is examined using the FEDScope. First, the mapping between module addresses and FED channels is reviewed. The readout speed of the TBM can be programmed to either 20 or 40 MHz. Changing the readout frequency of the TBM is well visible on the FEDScope and thus can be used to identify the FED channels corresponding to a given TBM address. Furthermore, the presence of a correctly digitized analog signal in the FED confirms the functioning of the trigger distribution, the token mechanism and the module readout sequence.

The procedure turned out to be well suited for a fast identification of broken detector modules since the malfunctioning of modules was in most cases due to wire-bonds which have been damaged during the handling. The corresponding failure modes, for instance the prevention of the token passage or the trigger distribution, could be spotted quickly when performing the testing sequence described above.

8.2 Performance Tests and Calibrations

More detailed performance tests and calibrations are performed using the CMS pixel online software implemented in the XDAQ framework. The following sequence of test and calibration stages is processed [7]:

1. Repeat the adjustment of the delay settings within the digital communication circuit.
2. Adjust the sampling point (delay and phase setting) in the FED for the digitization of the analog pulse.
3. Adjust the laser bias in the AOHs. The optimal setting is found by increasing the bias current until the separation between the ultra-black and the black level saturates. The AOH bias calibration is temperature dependent and has to be redone each time the operating temperature of the detector changes.
4. Set the ultra-black signals of the ROCs and TBMs to the correct amplitude [8]. The TBM and ROC ultra-black levels have to be readjusted in case the previous item changes.
5. Tune the offset in the FED to set the black level of the arriving analog signal to a predefined value in the middle of the ADC range. Due to the strong temperature dependence of the AOHs, the position of the black level baseline within the FED ADC range has to be monitored constantly and re-adjusted if the drift is too high.
6. Select a threshold and delay setting for each ROC in such a way that the injected test charge is correctly registered and readout. The amplitude of the injected signal is set by programming the corresponding DAC register (V_{cal}).
7. Perform the address level calibration. As discussed earlier, the pixel row and column address is encoded in 6 discrete analog levels which have to be well separated for being correctly decoded by the FED. The position of the six address levels is determined by measuring the levels of all pixels in a ROC and overlaying them in a histogram (Fig. 8.6). The decoding limits placed in the center between to neighboring peaks are then downloaded to the FED.
8. Run the pixel alive test. In this test, charges above threshold are injected into all pixels in a ROC and the correct response of each pixel is verified.
9. Check the high voltage connection of each module.
10. Determine the threshold and the noise of each pixel. This is done by measuring the so-called S-Curve which is the pixel response efficiency as a function of the injected charge. An S-Curve measurement recorded during module testing at PSI is shown in Fig. 8.7. The charge value corresponding to a response efficiency of 50% defines the threshold, while the noise is proportional to the width of the region where the efficiency changes from 0 to 100%. The S-Curve is measured for a subset of 81 pixels per ROC which was found to be sufficient to determine the average noise and threshold per ROC. The amplitude of the injected signal has been calibrated during the module tests using test data from X-ray sources of known energies [9]. An average slope of 65.5 electrons per DAC unit and an offset of -114 electrons was determined, the values change by up to 15% from ROC to ROC.
11. Calibrate the analog pulse height. An exact calibration of the pixel charge measurement is crucial for a precise position resolution since the hit position is interpolated from the charge information of all pixels in a cluster. The calibration is performed by injecting signals with increasing amplitudes to each pixel and measuring the analog pulse height. An example of a measurement of one pixel is shown in Fig. 8.8. For each pixel about 30 charge values are injected.

8.2 Performance Tests and Calibrations

Fig. 8.6 Address-levels of all pixels in a ROC as received by the FED. The lines are the separation limits used for the decoding of the pixel addresses in the FED

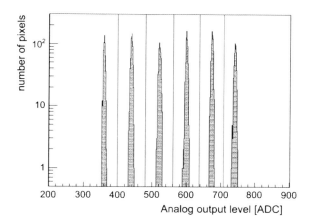

Fig. 8.7 Measurement of the pixel response efficiency as a function of the injected charge fitted with an error function (S-Curve). The value corresponding to a response efficiency of 50% determines the threshold, the noise is proportional to the width of the error function

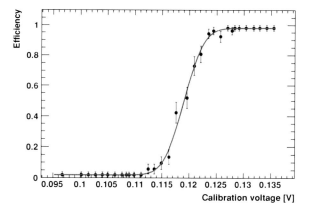

The pulse height curve is approximately linear below saturation at about 45, 000 electrons and can be parametrized by the slope (gain) and offset (pedestal) of a linear fit.

The result of these tests and calibrations at the different stages of the BPIX detector construction will be reviewed in the following.

8.3 Construction

The integration of the CMS BPIX detector took place between December 2007 and May 2008. The modules were mounted on the support structure at PSI while the assembly of the supply tubes was done at the University of Zürich. The final system was assembled and fully commissioned at PSI. Thereafter, the system was transported to CERN in a fully functional state, without disconnecting any components.

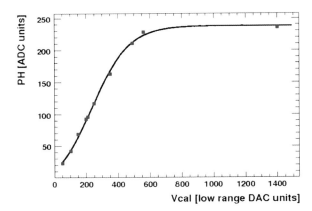

Fig. 8.8 Measured pulse height as a function of the amplitude of the injected signal. One V_{cal} DAC unit corresponds to 65.5 electrons

8.3.1 Module Mounting

The BPIX modules were built [10] and extensively tested at PSI [9, 11, 12]. The goal of the module tests at PSI was to verify that all pixels function correctly, each ROC can be programmed properly, and all calibrations of a module produce meaningful results. All modules were graded and only the best quality modules were used in the final system. In total, 948 modules have been manufactured and tested. The carbon BPIX detector mechanics was fabricated at the University of Zürich.

We mounted the 768 modules of the BPIX detector manually on the carbon support structure. The three layers of the support structure are divided into half cylinders and consist of 10, 16 and 22 ladders, respectively. The ladders are arranged alternating on the inner and the outer surface of the half cylinders and accommodate eight modules each. The first four modules were connected to the endring prints on the $-z$ side of the detector, the remaining four modules to the prints on the $+z$ side. The signal and power cable of the module had to be adjusted to match a particular module position on the mechanical structure. An elaborate procedure has been established to prepare the modules before mounting:

- Setting the TBM address by removing wire bonds according to a predefined address scheme.
- Cutting the Kapton signal cable to a precision of 0.5 mm compared to the predetermined length.
- Bending the signal cable with the help of a special bending tool that ensures a well defined bending radius and an exact position of the bending.
- Cutting the power cable with a precision of ≈ 2 mm.
- Soldering a plug to the shortened power cable.
- Attaching the power cable to the signal cable.

In order to protect the sensitive structures of the modules from mechanical stress and touches, the body of the module was stored in a box during all manipulations

8.3 Construction

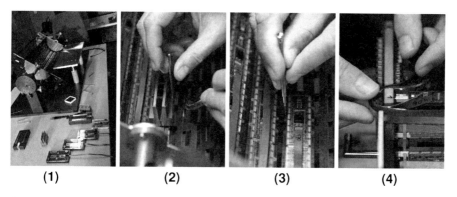

Fig. 8.9 Illustration of module mounting procedure: 1. Working place for the detector assembly. 2. Placing the module on the shell. 3. Screwing the module to the shell. 4. Placing cables at the end-flange

of the cables. Afterwards, the module was uncased and put on a jig equipped with two small pins that fit through the screw holes in the module base plate to keep the module in place. A dedicated mounting tool had been designed to facilitate the demanding task of lifting the module from the jig and placing it onto the ladder. The mounting tool is a clamp in which the module position is fixed with the aid of a 100 μm thin stainless steel sheet taking hold of the module base plate. It has a mechanical guidance to place the module on the ladder and two feedthroughs for the screwdriver. This allowed to screw the modules in a protected way onto the ladders. The module cables were fixed at the end of the ladder with a tiny cable clamp and connected to the PCB placed on the detector endflange. The mounting procedure is illustrated in Fig. 8.9.

The basic functionality of the modules was verified at different stages of the mounting procedure. This included a measurement of low and high voltage, a scanning of the module TBM addresses and a measurement of the analog readout levels. For the outermost layer these tests could be done sector-wise, once a readout group had been connected to its PCB. The PCB for the sectors of the first two layers however were mounted on the endflange of the third layer. Thus, the modules were tested individually after mounting and retested after the assembly of the three layers. In order to recover a broken module, the three layers had to be taken apart and the module and possibly also neighboring modules had to be disconnected and unscrewed. This was done several times and broken modules were either repaired or replaced.

The procedure proved very efficient. Up to 40 modules could be mounted in a day. Only three out of 768 modules were lost during the assembly, 10 modules were damaged but could be repaired and used in the final system. However, the replacement and repair of a broken module was an extremely delicate operation with a high risk of inducing more damage to the system. In the last iteration of testing at PSI, three modules (0.35%) were found to be non-working. Figure 8.10 shows a picture of one half of the fully assembled BPIX detector.

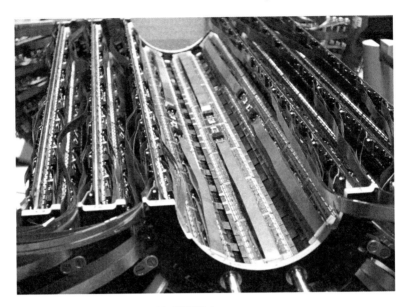

Fig. 8.10 One half of the fully assembled BPIX detector

When the BPIX detector was fully assembled a clearance test between the two half shells was performed. The test showed that the cables of the closely approaching half modules had to be slightly rearranged in order to provide enough clearance for the installation of the detector.

8.3.2 Supply Tube Assembly

The BPIX supply tube was fabricated at the University of Zürich between November 2007 and April 2008. The supporting elements of the supply tube structure are stainless steel cooling tubes running along the z-direction. They are connected to fiberglass stiffener rings and inner and outer aluminum flanges. The gaps in-between are filled with foamed material to guarantee the required rigidity. All power and slow control lines are embedded in the supply tube body. The BPIX supply tubes are equipped with a total of 124 temperature sensors and 8 humidity sensors. The temperature sensors are placed on the CCU boards, the AOH boards and on the supply tube cooling lines.

The CCU boards were produced at the University of Zürich and thoroughly tested at PSI using the System12 setup. The proper working of the CCU ring architecture and the redundancy mechanism was checked and the programming of the PLL and the DELAY25 chip was tested. With the first version of the boards problems occurred due to weak soldering joints in the PCB which easily broke when applying mechanical

8.3 Construction

Fig. 8.11 One of the four BPIX supply tubes during assembly placed on a rotatable mandril

stress, for instance by bending the boards. A redesign of the CCU board solved this problem. Furthermore, the temperature stability of the CCU boards was verified by successfully running them at $-20°C$ and $+55°C$. The boards proved to be reliable in long-term usage as they were operated in System12 for several months without any problems.

The detector readout electronics was integrated on the two halves of the supply tube in February and April, 2008. All components were tested for functionality before and during mounting. A standard PC with a FEC card was used to send the clock and the control commands to the CCUs, the returned digital signals were monitored on an oscilloscope and the output light intensity of the analog optical signals was measured.

In the first step, we mounted the CCU board and the analog and digital motherboards on the supply tube. We then placed the DOHs on the CCU board and connected them via optical fibers to the FEC card to establish communication and perform first functionality tests.

The mounting of the DOHs and AOHs and the inlaying of the optical fibers was done sector-wise wherefore the supply tube was placed on a rotatable mandril (Fig. 8.11). The main challenge was the arrangement of the 1,440 single optical fibers. For logistic reasons and for the purpose of flexibility during the assembly, all AOHs and DOHs are identical and the fibers glued to the boards all have a length of 2 m. The slack management of the fibers turned out to be very difficult as the space on the supply tube is limited and the bending radii of the fibers have to be more than 5 cm to guarantee maximum light transmission. At the far end of the supply tube optical connectors link the single fibers with fiber bundles of 12 fibers each. These bundles form pig tails with a length of about 1 m which then connect to the first patch panel inside CMS.

Fig. 8.12 Picture of an AOH where the two outer lasers are disconnected. The protection bar glued to the outer two lasers is well visible

We first mounted and tested the DOHs and then continued with the AOHs. A 40 MHz clock signal was injected into every DOH channel and the returned signal was checked with the help of an oscilloscope. Not a single non-working DOH channel was found. The plugging of the AOHs was more tedious as the wire-bond connections between the laser and the board were found to be very fragile. The problem was caused by a protection bar which is placed on top of the lasers and glued to the outer two lasers (see Fig. 8.12). This glued joint is unfortunately much stronger than the one between the lasers and the board. When bending the boards, the force is thus transferred to the wire-bonds which are likely to break. Fifteen AOHs (out of 192) had to be replaced during the assembly due to broken wire-bond connections of the laser.

The stability of the analog signal is strongly affected by the undesirable temperature dependence of the AOHs. The level of the analog signal is shifted by 50 ADC counts when the temperature of the AOH changes by 1°C [7]. The FED is able to internally correct for a drift within a temperature range of ±2°C. Consequently, an active cooling has to be provided to control the temperature of the AOHs within a very narrow range and assure a stable operation of the detector. For this reason, aluminum plates were placed on top of the AOHs and DOHs which connect them thermally to the supply tube cooling lines.

In a final step, the power and control cables for each sector and the central slot were mounted on the supply tube and the fibers were covered with a thin aluminum coated shielding.

8.3 Construction

Fig. 8.13 One half of the BPIX detector with the supply tubes connected. The detector system is placed on a rail system inside a 5 m long custom-built transport box

8.3.3 Integration of the Complete System

The detector and the supply tubes were assembled and commissioned at PSI within only two months in May and June 2008.

The detector halves together with the supply tubes were integrated in two 5 m long transport boxes custom-built for the installation into CMS. Within the transport box the detector is placed on wheels on a rail system which then can be used to slide the pixel detector inside CMS. A picture of the pixel detector with the two supply tubes inside the transport box can be seen in Fig. 8.13.

The connection between the readout sectors of the supply tube and the detector endflange is provided by four Kapton signal cables and a power cable. An aluminum clamp screwed to the detector endflange retains the Kapton cables in their position. The mechanical strength of the connection between the detector and the supply tube is determined by the stiffness of the signal and power cables.

In addition to the signal and power cables, silicon rubber hoses fixed with aluminum clips were mounted to connect the aluminum cooling lines of the detector mechanics and of the supply tubes. A leak test of the 10 cooling lines of each half detector was performed by filling the lines and monitoring the amount of coolant. No leaks were detected inside the pixel system.

After the assembly, the complete system was fully commissioned by performing the low level hardware tests followed by detailed functionality tests. The low level hardware tests were carried out using the standalone software and could be completed within less than one day while the detailed performance tests took about 3 h per sector.

The completely integrated system was disassembled several times for the repair or replacement of broken detector modules. In the end, the number of dead channels was found to be less than 0.4%. The dead channels were due to one module without high voltage connection, two modules with a broken token passage and one module with a bad ROC header. In addition, a sector without digital readback was found and

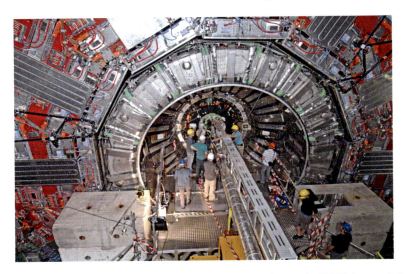

Fig. 8.14 Insertion of the BPIX detector inside CMS. In the picture, the BPIX detector is inside the transport box which is placed on the insertion table next to the beam pipe. The rail system inside CMS and the transport box are joint by extension rails

two broken AOH channels were observed. The former issue does not influence the analog readout and the latter was recovered by rerouting the signal through another channel.

8.4 Installation into CMS

The pixel detector was inserted into CMS after the installation and cabling of the silicon strip detector had been completed. The BPIX detector fits into the small volume limited by the outer radius of the beam pipe at 2.9 cm and the inner radius of the first layer of the strip tracker at about 21 cm. A system with bending rails on top and bottom inside CMS had been designed to insert the pixel detector and the supply tubes along the beam pipe. A clearance of 7–8 mm to the beam pipe had been calculated in simulations and checked with the help of a design model. The transport box with the pixel detector was placed on an insertion table and the rail system inside the box was joint with the rail system inside CMS using temporary extension rails (see Fig. 8.14). In this way, the pixel detector could slide out of the transport box into its final position. At the end, the service lines were connected at the so-called patch panel 0 (PP0) to the detector infrastructure.

The detector infrastructure in the CMS cavern had to be ready well in advance of the detector installation. This included the installation and commissioning of the power supply system and the detector safety system. The cables and fibers between the electronic racks and the detector were placed in winter and spring of 2008. The connectivity of all cables and fibers was tested and measurements of the signal transmission quality were performed. The FED and FEC modules located in the CMS service room had already been installed and tested at the end of 2007.

On April 25, 2008 the commissioning system was shipped to CERN and a test installation took place. The system was craned down to the cavern through the main shaft, lifted to the installation table and inserted into CMS. The installation of the commissioning system went smoothly and was finished within less than 4 h. The power cables and the optical fibers of the equipped sector were connected at the PP0 and the correct cable lengths were verified. The installation test did not reveal any need for mechanical adjustment before the final installation.

The two halves of the BPIX final detector system were transported to CERN on July 15, 2008. After the transport, we tested the system in the surface hall at P5 and no additional damage was found. The installation of the final system started on July 23. Both halves of the detector were lowered into the cavern on the same day. The insertion of the inner shell was completed without any problems and all the connections were made, a total of 40 power and control cables and 18 multi-fiber ribbons. In order to make a fast check-out possible a temporary cooling system was set up. The second half of the BPIX detector was inserted the following day. The first attempt failed due to a collision of the detector end-flanges of the two detector halves. This problem was solved by mechanically modifying the suspension of the insertion wheels to enlarge clearance between the two half shells, and allowed us to finalize the insertion successfully.

All power and control cables and all optical fibers of the BPIX detector were connected by July 24. A picture of the detector in the final position and the connection area PP0 is shown in Fig. 8.15. The BPIX services had to be disconnected again when the final cooling tubes were joined and when the forward pixel detector was installed.

The cooling lines of the pixel supply tube are connected to the cooling system using stainless steel flexible pipes. In the first running period after the installation the cooling fluid had a temperature of $+17°C$ and the 10 barrel pixel cooling loops did not show any leak.

8.5 Commissioning and Performance

The testing of the pixel system started as soon as the pixel services were connected and cooling was available. The low level hardware tests were done first to quickly evaluate the performance of all detector components. On July 29, all sectors of the BPIX detector were working satisfactorily so that the construction of CMS could proceed with the installation of the forward pixel detector. On August 7, the closing of the CMS detector started and thus access to the pixel detector was no longer possible. A period of extensively testing and calibrating the detector before the first cosmic data taking took place in October and November 2008. The results of the testing and the performance of the BPIX detector are reviewed in the following.

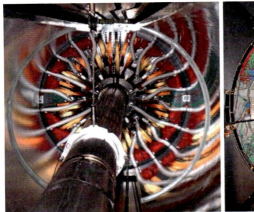 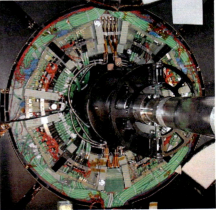

Fig. 8.15 View of the BPIX detector in its final position. *Left* The picture shows the BPIX detector inside CMS enclosing the central beam pipe. The camera is placed inside the detector supply tube and the detector endflange covered by the cooling lines and signal and power cables can be seen. *Right* Picture of the connection area PP0 at the far end of the pixel volume. The connection of power cables (*green cables*), optical fibers (wrapped in *red rubber bands*) and cooling lines are visible. In the foreground the beam pipe suspension is shown

8.5.1 Performance of the Optical Links

The optical signals pass four connection points before they are translated into electronic signals in the FED modules in the service room: a single-fiber (MU) connection on the supply tube located at 2 m distance from the laser, a multi-fiber (ribbon) connection at the pixel detector PP0 at a distance of 3 m, a multi-ribbon connection at the strip tracker PP1 at 7 m distance and a connection to the FED modules at 63 m distance. The latter connection at the strip tracker end-flange was not accessible anymore at the time of the BPIX installation.

The digital-optical ring is tested by sending and receiving a 40 MHz clock signal. For the analog-optical lines the scan of the laser bias range was repeated. After three iterations of re-cleaning the PP0 connections, the optical fibers for the transmission of the digital signal showed an excellent performance while 29 out of 96 ribbons of the analog readout contained noisy fibers. These 29 ribbons were investigated with the help of an optical reflectometer (OTDR) and a visual inspection of the connection to the FED. The OTDR measurement did not show any reflection at the PP0 connections which means that an optimum light transmission was provided. However, in 19 cases a reflection peak at a MU connection was spotted. As the MU connectors are located on the supply tube, it was not possible to improve the connection by cleaning. In 11 cases this reflection influences either the noise or the slope of the calibration curve for the corresponding channel. Fortunately, further testing indicates that this does not degrade the performance of the analog address level decoding substantially.

8.5 Commissioning and Performance

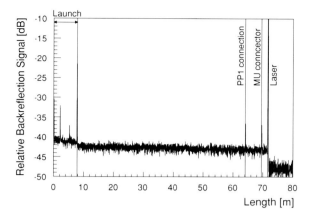

Fig. 8.16 OTDR measurement of a fiber with a reflection peak at the MU and the PP1 connection

In addition, a bad PP1 connection was found. The four fibers passing through that connection could be recovered by using a spare ribbon. An example of the OTDR measurement of a fiber with a reflection at the MU and the PP1 connection is shown in Fig. 8.16.

The visual inspection of the connection to the FED was performed with a microscope and 19 ribbons with marks on the fibers were spotted. In about 50% of the cases cleaning was successful and slightly improved the noise behavior. A microscopic view of the FED connection with marks on the fibers can be seen in Fig. 8.17.

Not a single optical fiber of the BPIX detector was lost during all operations.

8.5.2 Detector Module Functionality

Although the testing after the transport of the BPIX detector has not revealed any damage (in addition to the four modules known to be non-working), after the installation five more broken modules were identified. The failure modes are:

- 3 modules without high-voltage connection,
- 1 module with a bad ROC header,
- 1 module that could not be programmed.

Furthermore, 4 individual ROCs did not produce valid signals. An additional module with a bad TBM header was recovered by rerouting the signal through the other TBM. In total 100 ROCs (0.87%) of the BPIX detector could not be operated, 80% of the failure modes were explained by broken wire bonds or missing high voltage connection, the remaining 20% were due to modules which did not respond to programming and thus had to be disabled. The number of dead pixels on otherwise functional ROCs is very low (0.01%). These failures are caused by faulty bump-bond connections between the ROC and the sensor and have already been observed during module testing.

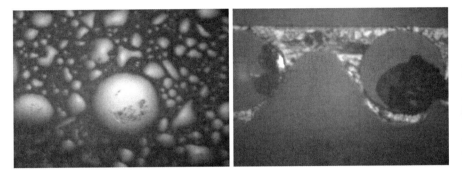

Fig. 8.17 Microscope picture of the optical connection at the FED. On the *left hand side* a single optical fiber is shown, while on the *right hand side* two optical lines inside the FED connector are visible. All fibers are partially covered by dirt

8.5.3 Results of the Detector Calibration

In this section the results of the BPIX detector testing and calibration period in summer and autumn 2008 are summarized. More details can be found in [13]. The sequence of tests introduced in Sect. 8.2 were performed. After determining a working region for each module and adjusting the analog levels in the AOHs, front-end chips and FED channels, an address level calibration for each ROC was run. The quality of the address encoding is evaluated by comparing the width of the address peaks to the separation between to neighboring pixels. The result is shown in Fig. 8.18. The level separation is considerably larger than the width of the peaks, even in comparison with the broadest peaks. The few channels affected by dirty optical connectors have a smaller level separation, however still large enough for a reliable address decoding. The address level calibration is temperature dependent and has to be repeated every time the operating conditions change.

The threshold values for the BPIX modules were programmed to the target values determined in the module testing. Afterwards the threshold and the noise were measured using the S-Curve method. An average threshold of 3,829 electrons was found, well above the noise level at 141 electrons.

The results of the analog pulse height calibration are given in Fig. 8.19. The gain and pedestals are obtained from a linear fit. The tail in the pedestal distribution is due to misconfiguration of individual ROCs which was recovered at a later stage of detector calibration.

8.5.4 Results of the Cosmic Run

The goal of the cosmic data taking with the CMS detector is the commissioning and calibration of the individual subdetectors, the alignment of the tracking detectors and the muon chambers and the testing of the DAQ system.

8.5 Commissioning and Performance

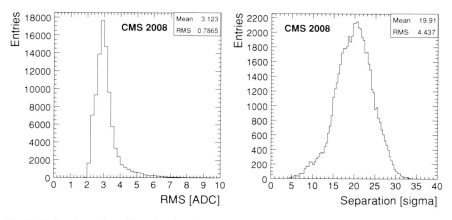

Fig. 8.18 Results of the address level calibration [13]. *Left* RMS of width of the six address level peaks for all operable ROCs in the detector. *Right* Separation between the mean of two neighboring peaks. The separation is given in units of sigma, defined by summing in quadrature the RMS widths for adjacent peaks

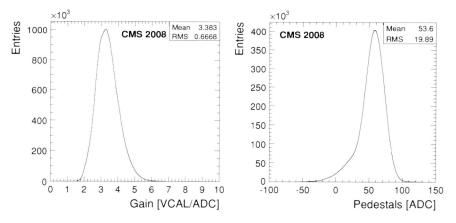

Fig. 8.19 Distribution of gain (*left*) and pedestals (*right*) for all pixels extracted from the linear fit to the pulse height curve [13]

The pixel detector showed a very good performance during cosmic data taking and proved its ability for stable running. The main cause of interruption during the pixel detector operation was problems with the cooling plant or the power supply. Together with some rare failures of the detector control software this amounted to an overall data taking efficiency of 97%.

The CMS experiment recorded about 270 million cosmic-ray triggered events with the solenoid at a field strength of 3.8 T. More than 4 million tracks were reconstructed out of which approximately 85,000 did cross the pixel detector volume. An event display showing a cosmic muon traversing the pixel detector is shown in Fig. 8.20.

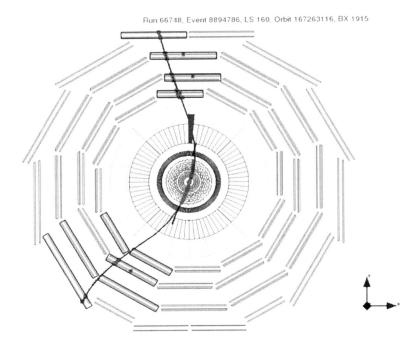

Fig. 8.20 CMS event display showing a cosmic muon traversing the pixel detector

The results of the alignment of the silicon strip and pixel detectors is detailed in [14]. After performing a track-based alignment, the precision of the detector position with respect to particle trajectories had been derived from the distribution of the median of the cosmic muon track residuals measured in each module. A precision of 3 μm in the $r\phi$ direction and 4 μm in the z direction has been achieved for the BPIX detector.

In total about 257, 000 hits were reconstructed in the pixel detector with an average number of 60 hits per ROC. In Fig. 8.21 the number of track-associated hits in each BPIX ROC is shown. The ROCs that did not have any hits were excluded from the readout for the reasons given in Sect. 8.5.2.

Furthermore, the cosmic data were used to estimate the hit efficiency of the BPIX modules from data. This is done by extrapolating tracks reconstructed in the silicon tracker to the pixel detector and checking the presence of a compatible hit. The efficiency is then defined as the number of tracks with a compatible pixel hit divided by the total number of tracks. For the efficiency measurement only tracks to which one additional hit was associated in both the top and bottom half of the pixel detector were selected. A layer efficiency averaged over the operable modules of (97.1±1.4)%, (97.1±1.9)% and (96.4±2.6)% was measured for the first, second, and third barrel layers, respectively.

8.6 Summary

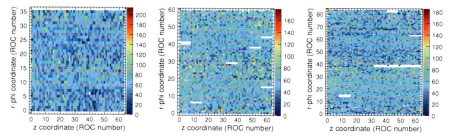

Fig. 8.21 Number of hits associated to a track detected in each ROC for the first (*left*), second (*middle*) and third (*right*) barrel layers [13]. Bins in white correspond to readout chips excluded from data taking

8.6 Summary

The contributions to the construction and commissioning of the CMS pixel barrel detector made during this thesis were presented in this chapter. The construction of the BPIX detector included the mounting of the modules on the mechanical cooling structure, the assembly of the supply tube and the integration of the complete system.

A slice of the CMS control and data acquisition system was established at PSI in order to gain experience in operating the complete system and fully commission the BPIX detector before transporting it to CERN. A sophisticated testing procedure has been developed and implemented in a standalone software framework. This procedure allows to verify the functioning of the main detector components in a short period of time and was therefore of utmost importance for the testing phase after the installation at CERN.

The installation was completed within only a few days and the first test of the BPIX detector revealed an excellent performance with less than 1% dead channels.

After a period of extensively testing and calibrating the pixel detector at CERN, the first data taking of cosmic data took place in October and November, 2008. During this period the detector was running stable with a high data taking efficiency.

References

1. L. Caminada, A. Starodumov, Building and commissioning of the CMS pixel barrel detector. JINST **4**, P03017 (2009)
2. S. Sechrest, *An Introductory 4.4.bsd Interprocess Communication Tutorial* (Computer Science Research Group, University of California, Berkeley, 1993)
3. Python Programming Language, http://www.python.org
4. J. Gutleber, L. Orsini, Software architecture for processing clusters based on $I_2 O$. Clust. Comput. **5**, 55 (2002)
5. ROOT: An object oriented data analysis framework, http://root.cern.ch
6. P. Placidi et al., CMS tracker PLL reference manual. CERN Document Server
7. D. Kotlinski, Status of the CMS pixel detector. JINST **4**, P03019 (2009)

8. H.C. Kästli et al., Design and performance of the CMS pixel detector readout chip. Nucl. Instrum. Meth. A **565**, 188 (2006)
9. P. Trüb, CMS pixel module qualification and Monte-Carlo study of $H \rightarrow \tau^+\tau^- \rightarrow l^+l^- \not{E}_T$. Ph.D. Thesis, ETH Zürich, ETH No. 17985, 2008
10. S. König et al., Building CMS pixel barrel detector modules. Nucl. Instrum. Meth. A **582**, 776 (2007)
11. C. Eggel, CMS pixel module qualification and search for $B_s^0 \rightarrow \mu^+\mu^-$. Ph.D. Thesis, ETH Zürich, ETH No. 18232, 2009
12. A. Starodumov et al., Qualification procedures of the CMS pixel barrel modules. Nucl. Instrum. Meth. A **565**, 67 (2006)
13. CMS Collaboration, Commissioning and performance of the CMS pixel tracker with cosmic ray muons. JINST **5**, T03007 (2010)
14. CMS Collaboration, Alignment of the CMS silicon tracker during commissioning with cosmic rays. JINST **5**, T03009 (2010)

Chapter 9
Conclusion and Outlook

A study of the inclusive b-quark production at the CMS experiment has been presented within this work. Thanks to the large b-quark production cross section at the LHC, high statistics data samples are available soon after the LHC startup. The measurement of the b-quark production cross section with these data is therefore a prime candidate to yield one of the first physics result obtained from proton-proton collisions at a center-of-mass energy of $\sqrt{s} = 7$ TeV.

As a result of this thesis, an analysis strategy has been developed which focusses on the reconstruction of muons originating from semileptonic decays of b-quarks. The fraction of signal events in data is determined on a statistical basis by performing a fit to the measured $p_\perp^{\text{rel}}$ distribution by means of simulated templates for signal and background events.

The $p_\perp^{\text{rel}}$ variable is defined with respect to the fragmentation jet direction reconstructed from charged particles tracks only. With this technique an efficient reconstruction of the jet direction is possible even for very low-energy jets.

The event selection and the performance of the physics object reconstruction has been studied in collision data and simulation and in general a remarkably good agreement is found.

A first measurement of the inclusive b-quark production cross-section at a center-of-mass energy of $\sqrt{s} = 7$ TeV has been presented. The measurement is based on data statistics corresponding to an integrated luminosity of $\mathcal{L} = 8.1$ nb^{-1}. The data has been recorded by the CMS experiment during the first months of data taking in April and May, 2010.

The preliminary result for the total inclusive b-quark production cross-section in the visible kinematic range is

$$\sigma(pp \to b + X \to \mu + X', p_T^\mu > 6 \text{ GeV}, |\eta^\mu| < 2.1) \tag{9.1}$$
$$= (1.48 \pm 0.04_{\text{stat}} \pm 0.22_{\text{syst}} \pm 0.16_{\text{lumi}}) \, \mu\text{b}.$$

Furthermore, a measurement of the differential b-quark production cross-section as a function of muon transverse momentum and pseudorapidity was performed. The

---

L. Caminada, *Study of the Inclusive Beauty Production at CMS and Construction and Commissioning of the CMS Pixel Barrel Detector*, Springer Theses, DOI: 10.1007/978-3-642-24562-6_9, © Springer-Verlag Berlin Heidelberg 2012

measurement constitutes a sensitive probe of the predictions of perturbative QCD at the unprecedented high energy scale provided by LHC. It was compared to the leading order and next-to-leading order MC predictions. The data tends to be higher than the MC@NLO prediction at low transverse momentum and central rapidity.

In the second part of this thesis, the hardware related work has been presented. The integration of the CMS pixel barrel detector has been accomplished within about two years. The availability of a test stand at PSI has proven particularly important for commissioning the individual detector components as well as for operating the final detector system. It allowed to transport the detector in a fully functional state to CERN and to install it into CMS within the tight schedule.

A tremendous joint effort in commissioning and calibrating resulted in the successful and stable operation of the CMS pixel detector during cosmic and collision data taking.

The CMS pixel detector has demonstrated excellent performance and is going to play a key role in tracking based physics analyses like the b-quark production measurement presented in this work.

Appendix A: Maximum Likelihood Fits

A.1 Bins in Muon Transverse Momentum

A.1.1 Fit Result

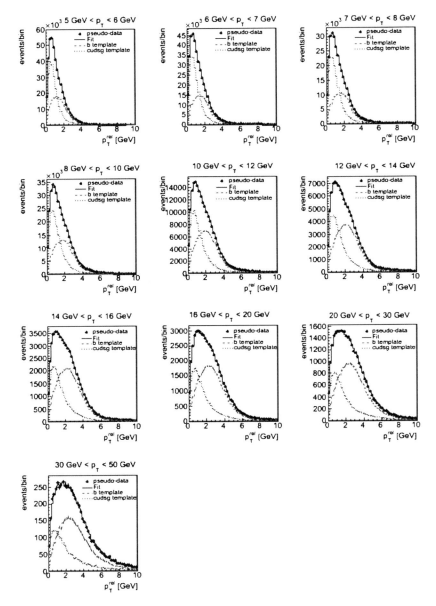

Fig. A.1 Fit result obtained by dividing the MC sample into two independent subsamples and using the approximate fitting method. The *dashed* and the *dotted line* are the *b*- and *cudsg*-template, respectively. The *full circles* correspond to the data distribution, while the *solid line* is the result of the fitting procedure. The "data" distribution is scaled to an integrated luminosity of 1 pb^{-1}. All bins in muon transverse momentum are shown

Appendix A: Maximum Likelihood Fits

A.1.2 Fit Deviation in the Approximate Method

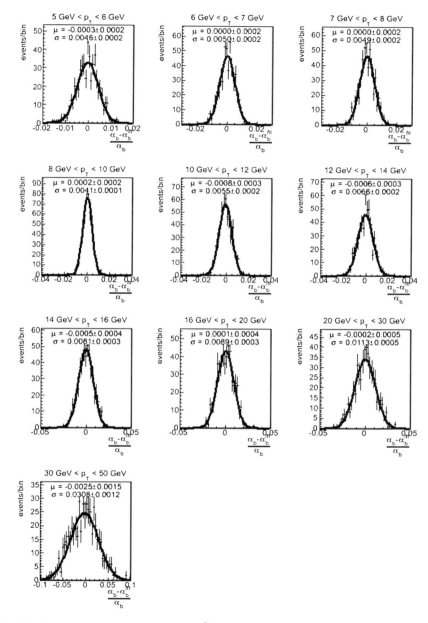

Fig. A.2 Deviation of the fitted scale factor α_b^{fit} in b-events from the true value α_b, when using the approximate fitting method. The results are obtained from repeated pseudo experiments with data statistics generated by appropriate random variations corresponding to an integrated luminosity of 1 pb^{-1}. All bins in muon transverse momentum are shown

A.1.3 Pull Distributions of the Approximate Method

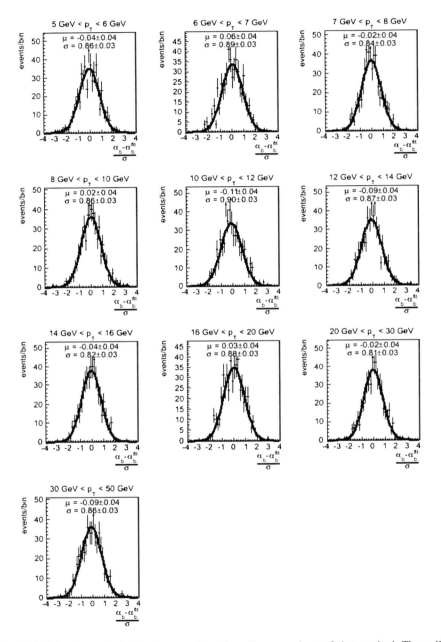

Fig. A.3 *b*-fraction pull distributions resulting from the approximate fitting method. The pull distributions are obtained from repeated pseudo experiments with data statistics generated by appropriate random variations corresponding to an integrated luminosity of 1 pb^{-1}. All bins in muon transverse momentum are shown

Appendix A: Maximum Likelihood Fits 143

A.1.4 Fit Deviation in the Full Treatment

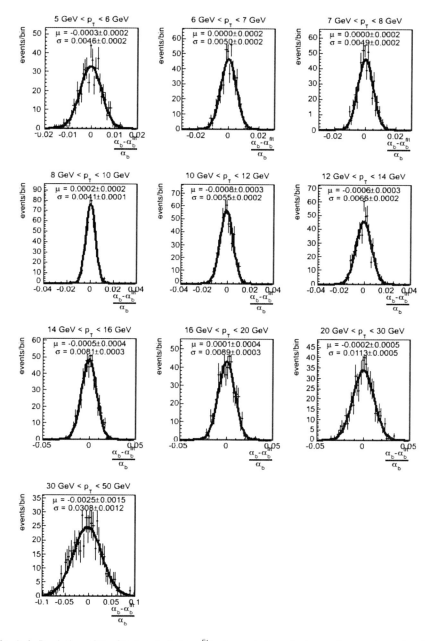

Fig. A.4 Deviation of the fitted scale factor α_b^{fit} in b-events from the true value α_b when using the full treatment in the fitting procedure. The results are obtained from repeated pseudo experiments with data statistics generated by appropriate random variations corresponding to an integrated luminosity of 1 pb^{-1}. All bins in muon transverse momentum are shown

A.1.5 Pull Distributions of the Full Treatment

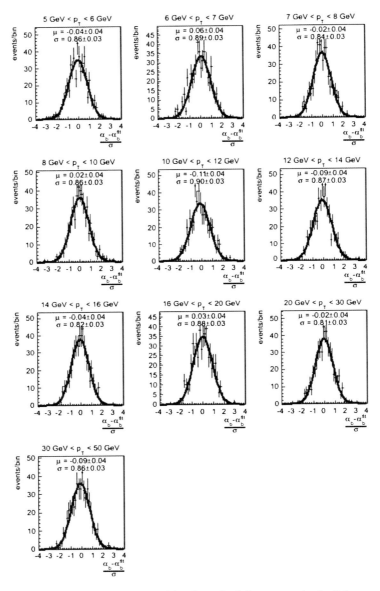

Fig. A.5 *b*-fraction pull distributions resulting from the full treatment in the fitting procedure. The pull distributions are obtained from repeated pseudo experiments with data statistics generated by appropriate random variations corresponding to an integrated luminosity of 1 pb^{-1}. All bins in muon transverse momentum are shown

Appendix A: Maximum Likelihood Fits 145

A.2 Bins in Muon Pseudorapidity

A.2.1 Fit Result

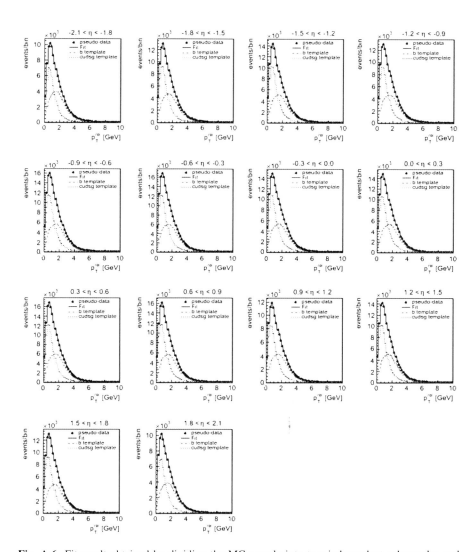

Fig. A.6 Fit result obtained by dividing the MC sample into two independent subsamples and using the approximate fitting method. The *dashed* and the *dotted line* are the *b*-and *cudsg*-template, respectively. The *full circles* correspond to the data distribution, while the *solid line* is the result of the fitting procedure. The "data" distribution is scaled to an integrated luminosity of 1 pb^{-1}. All bins in muon pseudorapidity are shown

A.2.2 Fit Deviation in the Approximate Method

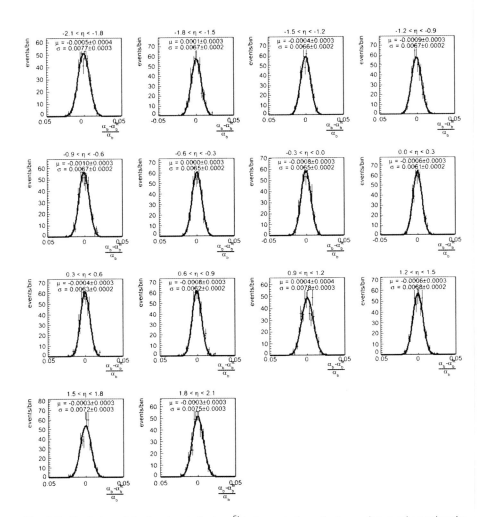

Fig. A.7 Deviation of the fitted scale factor α_b^{fit} in b-events from the true value α_b when using the approximate fitting method. The results are obtained from repeated pseudo experiments with data statistics generated by appropriate random variations corresponding to an integrated luminosity of 1 pb^{-1}. All bins in muon pseudorapidity are shown

Appendix A: Maximum Likelihood Fits 147

A.2.3 Pull Distributions of the Approximate Method

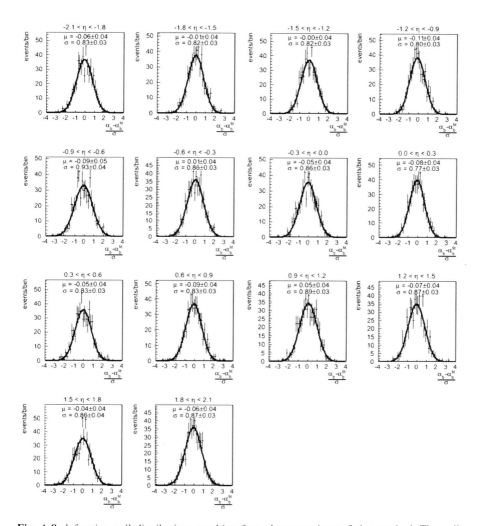

Fig. A.8 b-fraction pull distributions resulting from the approximate fitting method. The pull distributions are obtained from repeated pseudo experiments with data statistics generated by appropriate random variations corresponding to an integrated luminosity of 1 pb^{-1}. All bins in muon pseudorapidity are shown

A.2.4 Fit Deviation in the Full Treatment

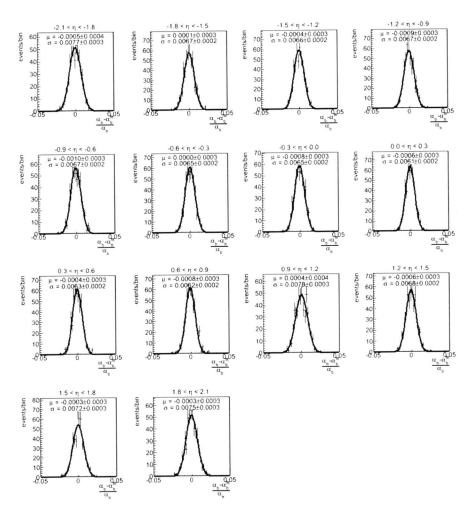

Fig. A.9 Deviation of the fitted scale factor α_b^{fit} in b-events from the true value α_b when using the full treatment in the fitting procedure. The results are obtained from repeated pseudo experiments with data statistics generated by appropriate random variations corresponding to an integrated luminosity of 1 pb^{-1}. All bins in muon pseudorapidity are shown

Appendix A: Maximum Likelihood Fits

A.2.5 Pull Distributions of the Full Treatment

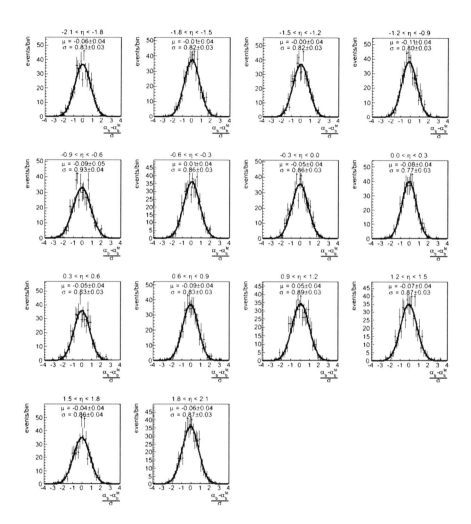

Fig. A.10 b-fraction pull distributions resulting from the full treatment in the fitting procedure. The pull distributions are obtained from repeated pseudo experiments with data statistics generated by appropriate random variations corresponding to an integrated luminosity of 1 pb^{-1}. All bins in muon pseudorapidity are shown

Appendix B: Systematic Uncertainties

B.1 Limited MC Statistics

Table B.1 Systematic uncertainties due to limited MC and data statistics

Relative error of fitted b-fraction ($\mathcal{L} = 1$ pb^{-1})	
p_T^μ	
5–6 GeV	0.5%
6–7 GeV	0.5%
7–8 GeV	0.5%
8–10 GeV	0.5%
10–12 GeV	0.6%
12–14 GeV	0.6%
14–16 GeV	1%
16–20 GeV	1%
20–30 GeV	1.2%
30–50 GeV	3%
η^μ	
$(-2, -1.8)$	0.8%
$(-1.8, -1.5)$	0.7%
$(-1.5, -1.2)$	0.7%
$(-1.2, -0.9)$	0.7%
$(-0.9, -0.6)$	0.7%
$(-0.6, -0.3)$	0.7%
$(0.3, 0)$	0.7%
$(0, 0.3)$	0.7%
$(0.3, 0.6)$	0.7%
$(0.6, 0.9)$	0.7%
$(0.9, 1.2)$	0.7%
$(1.2, 1.5)$	0.7%
$(1.5, 1.8)$	0.7%
$(1.8, 2.1)$	0.8%

L. Caminada, *Study of the Inclusive Beauty Production at CMS and Construction and Commissioning of the CMS Pixel Barrel Detector*, Springer Theses, DOI: 10.1007/978-3-642-24562-6, © Springer-Verlag Berlin Heidelberg 2012

Appendix C: Preliminary Results of First Collisions at $\sqrt{s} = 7$ TeV

C.1 $p_\perp^{rel}$ Distribution in Bins of Muon Transverse Momentum

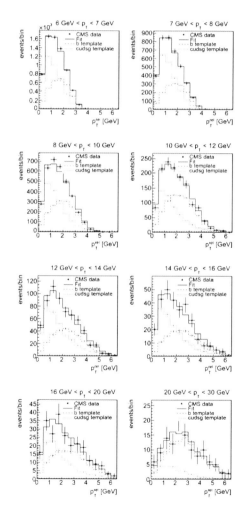

Fig. C.1 $p_\perp^{rel}$ distribution measured in data. Eight bins in muon transverse momentum are shown. The result of the maximum likelihood fit and the simulated template distributions are overlaid on the $p_\perp^{rel}$ distribution in data. The *dashed* and the *dotted* line are the *b*- and *cudsg*-template, respectively. The *full circles* correspond to the data distribution, while the *solid line* is the result of the fitting procedure

C.2 $p_\perp^{rel}$ Distribution in Bins of Muon Pseudorapidity

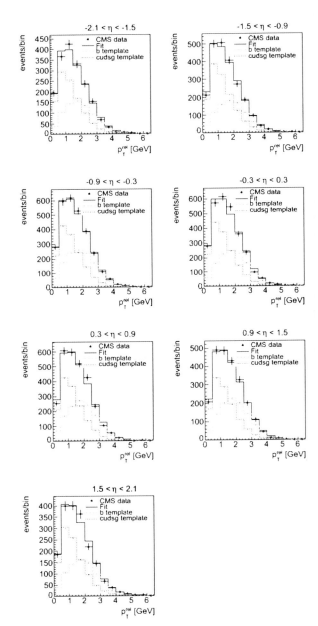

Fig. C.2 $p_\perp^{rel}$ distribution measured in data. Seven bins in muon pseudorapidity are shown. The result of the maximum likelihood fit and the simulated template distributions are overlaid on the $p_\perp^{rel}$ distribution in data. The *dashed* and the *dotted line* are the *b*- and *cudsg*-template, respectively. The *full circles* correspond to the data distribution, while the *solid line* is the result of the fitting procedure

Curriculum Vitae

Personal Data

Name: Lea Michaela Caminada
Date of Birth: December 30, 1981
Place of Birth: Zürich (ZH)
Nationality: Swiss
Citizen of: Vrin (GR) and Zürich (ZH)

Education

2006–2010	Doctoral studies at ETH Zürich and PSI Research at the CMS experiment at the CERN LHC
2006	Diploma in Physics at ETH Zürich summa cum laude Diploma Thesis at the H1 experiment at DESY in Hamburg in the group of Prof. R. Eichler *Title: Implementation of a Trigger for the Decay $b \rightarrow e + X$ on the Third Trigger Level at the H1 experiment*
2003–2004	Exchange year at the Strathclyde University in Glasgow
2001–2006	Studies of Physics at ETH Zürich
2001	Eidgenössische Matura Typus B
1994–2001	Gymnasium in Zürich-Oerlikon
1988–1994	Primary School in Zürich-Höngg

Awards

2010	CMS Thesis Award
2010	*Medallie der ETH Zürich* for an excellent PhD thesis
2006	Diploma in Physics at ETH Zürich summ cum laude

Printed by Publishers' Graphics LLC USA
SO20120319-061
2012